The Internet of Women^{sm}
Accelerating Culture Change

The Internet of Women^{sm}
Accelerating Culture Change

Editors

Nada Anid (Co-author and editor)
New York Institute of Technology, USA

Laurie Cantileno (Co-author and editor)
Cisco

Monique J. Morrow (Co-editor)
Cisco

Rahilla Zafar (Lead author and editor)
ConsenSys

LONDON AND NEW YORK

Published 2016 by River Publishers
River Publishers
Alsbjergvej 10, 9260 Gistrup, Denmark
www.riverpublishers.com

Distributed exclusively by Routledge
4 Park Square, Milton Park, Abingdon, Oxon OX14 4RN
605 Third Avenue, New York, NY 10158

First published in paperback 2024

The Internet of Women^{sm} Accelerating Culture Change / by Nada Anid, Laurie Cantileno, Monique J. Morrow, Rahilla Zafar.

© 2016 River Publishers. All rights reserved. No part of this publication may be reproduced, stored in a retrieval systems, or transmitted in any form or by any means, mechanical, photocopying, recording or otherwise, without prior written permission of the publishers.

Routledge is an imprint of the Taylor & Francis Group, an informa business

Publisher's Note
The publisher has gone to great lengths to ensure the quality of this reprint but points out that some imperfections in the original copies may be apparent.

While every effort is made to provide dependable information, the publisher, authors, and editors cannot be held responsible for any errors or omissions.

ISBN: 978-87-93379-68-8 (hbk)
ISBN: 978-87-7004-455-4 (pbk)
ISBN: 978-1-003-33973-1 (ebk)

DOI: 10.1201/9781003339731

Contents

Dedication ix

Endorsements xi

Foreword xv

PART I: Millennials Leading: Exploring Challenges and Opportunities Facing the Next Generation of Women in Technology

1 Building Communities through Technology   **3**
   1.1 Introduction 3
   1.2 How Shiza Shahid Is Supporting Female Founders and Their Critical Role in Reshaping the World 3
   1.3 An Activist's Evolution into Banking and Technology 7
   1.4 Young People Have Power: Millennials and STEM 13
   1.5 Launching a Female-Inclusive Tech Sector in Gaza 19
   1.6 Going beyond Gender and Tackling "Diversity Debt" in Technology 26
   1.7 From MIT to Supporting Women and Internet Penetration in Jamaica 30
   1.8 Gender Balance and Technology in the UAE 37
   1.9 Millennials Moving the Needle on Gender Equality 45
   1.10 Supporting Women in Embracing Technology in Kurdistan 48

PART II: Men and Women Empowering One Another

2 Behind Every Great Woman, There May Be a Man — 55
- 2.1 Introduction — 55
- 2.2 Supportive Strategies for Women Employees: Lessons Learned from Zambia — 55
- 2.3 The Necessity of Empowering Women — 60
- 2.4 How to Attract (and Keep!) Mothers in Your Company — 62
- 2.5 NA3M We Can: A Saudi Prince Roars for Women Gamers — 67
- 2.6 Female Leadership and the Future of Water Security in the Middle East — 69
- 2.7 Shifting the Gender Paradigm in Saudi Arabia: Thousands of Women Enter the Workforce — 73
- 2.8 Governments, Corporations, and Paying Women to Learn — 78

PART III: Bold Leadership: Women Changing the Culture of Investment and Entrepreneurship

Introduction — 85

3 Targeting Untapped Markets — 87
- 3.1 Creating an Ecosystem for Black Founders — 87
- 3.2 Solving the Technology Skills Gap through Investing in Minority Communities — 89
- 3.3 The Global Opportunity for Fintech Female Founders — 93

4 Mentoring Investors and Entrepreneurs — 99
- 4.1 Supporting Female Entrepreneurs and Creating Role Models in Israel and Beyond — 99
- 4.2 Female Entrepreneurs Making the Connection — 103
- 4.3 The Importance of Finding "Champions" for Entrepreneurs — 105
- 4.4 Bringing Water to the Women's Capital Desert — 110
- 4.5 Think Big: Expanding Plus-Sized Fashion Globally — 115

PART IV: Education for the 21st Century

Introduction — 121

5 Women in Academia: A Potential STEM Powerhouse — 123
- 5.1 What a Gender-Blind STEM World Could Look Like — 123
- 5.2 What Is at Stake? — 124
- 5.3 The Gateway to Opportunity: Academia in Flux — 129
- 5.4 What Are Women up Against? — 131
- 5.5 Strategies for Closing Academia's STEM Gender Gap — 142
- 5.6 Academia's Own Learning Curve — 148

6 Reshaping Traditional Ways of Learning — 157
- 6.1 Unearthing the Magic of STEAM — 157
- 6.2 Changing University Education through the Internet of Things — 162
- 6.3 Education and "Being a Girl" — 165
- 6.4 Creating a Culture of Peace and Understanding through Technology Education — 170
- 6.5 Internet of Women Pakistan: Divided by Access or Skills? — 176
- 6.6 Bridging the Digital Gap in Europe — 184

7 Educating Women in Post-Conflict Zones — 197
- 7.1 Developing a Technology Community for Women in Afghanistan — 197
- 7.2 Overcoming the Wounds of Genocide in Rwanda through Education and Technology — 200
- 7.3 How Technology Is Connecting Communities of Women in the DRC — 205
- 7.4 Inviting Multinationals to Support Youth and Technology in Gaza — 211

8 Opportunities for Adult Learners — 217
- 8.1 Queen Rania's Initiative to Provide High-Quality Education through Online Learning — 217
- 8.2 Creating a Developer Network for IT and Network Professionals and Software and Hardware Developers — 219
- 8.3 Are We Too Old to Be of Value? — 223
- 8.4 Open Source: Shifting the Corporate Innovation Model (and Leveling the Playing Field) — 229

PART V: Breaking the Glass Ceiling: A Generation of Women Forging ahead into Technology Leadership

9 Stories of Resilience, Perseverance, and Staying on Top of Technology Trends — 235
- 9.1 Introduction — 235
- 9.2 Breaking the Glass Ceiling in Saudi Arabia — 235
- 9.3 Building the Foundation for Innovation — 239
- 9.4 Career Advancement through Staying at the Forefront of Technology — 244
- 9.5 Being a Woman on an All-Male Executive Board — 248
- 9.6 Gender Bias in the Tech Sector — 251
- 9.7 A Snapshot of Women in Chinese History — 253

PART VI: Emerging Fields of Technology

Introduction — 261

10 Defining the Cutting Edge — 263
- 10.1 Meet the Woman Who's Created the 21st-Century Finance Model for Emerging Technologies — 263
- 10.2 Women Leading the World of Blockchain — 269
- 10.3 To Be or To Build: Women and Artificial Intelligence — 276
- 10.4 Opportunities for Women in the Green Economy and Digital Infrastructure — 281
- 10.5 CyberSheroes: The Inspiration Behind CSI — 290
- 10.6 Women in Cyber: Filling the Gap — 293

Manifesto — 303

Closing — 311

Afterword — 323

List of Contributors — 335

Rahilla Zafar

I dedicate this book to my father, Dr. Muhammad Zafar, a gentleman and a true champion for all women. A very special thanks to all my friends and colleagues who contributed so generously to this book, their stories and quotes are throughout these pages.

I'd also like to thank Sam Cassatt, Kate Rothschild, Dino Angaritis, Lina Lazaar, Sarah Ghaleb, Mina Sharif, Kelly Peeler, Nafez Dakkak, Areej Huniti, Ala Ebtekar, Seema Khan, Yasmine Rasool, Alaa' Odeh, Sana Odeh, Logan Shafer, Alex Klokus, Amanda Gutterman, John Lilic, Omar Christidis, Nina Curley, Nafeesa Syeed, Musa Syeed, Yasmin Altwaijri, Jon Lee Anderson, Michelle Anderson Binczak, Shannon Berning, Mukul Pandya, Marvin Mathew, Emily Ingebretsen, Jonathan Olinger, and Rummana Hussain.

My siblings: Azam, Aqila, and Jamila who are the most interesting and delightful people I know and my niece Lara Rahman who's now leader of the pack. Lastly, I'd like to thank my mother Nasim Zafar for all but forcing me to not live in a comfort zone and explore the world and even giving me a name that means traveler in Arabic.

Nada Anid

*To my daughters. . . my heroes. . . my inspiration. . .
To Ingrid and Audrée*

Laurie Cantileno

I dedicate this book to my husband Thomas J. Lillis who continually supports all my efforts and adventures no matter what, to my father Frank Cantileno who always told me I am capable of anything, my mother Shirley V. Oakman who's biggest dream in life was to be a writer and always gave me her utmost confidence and love. Additionally, to Phil Baker and Don Neault of Cisco executive management who's continual day to day support is unmatched, my mentor Eileen Westerkamp for her sponsorship over the years, and lastly to Todd Bowden who started me on this journey.

Monique J. Morrow

This book is dedicated to my mother, Odette G. Morrow, who passed away on September 17, 2013; to my father, Samuel A. Morrow Sr.; Veronique Thevenaz; Andre C. Morrow; Samuel A. Morrow Jr.; and Michelle M. Kline. You have all been my shining lights! Thank you for your love and encouragement.

I would also like to thank my co-Editors, Rahilla, Nada and Laurie for their valuable insights in developing this book, which is now a global movement!

To all our chapter contributors, we thank you for sharing your insightful narratives and recommendations to accelerate the change we all desire globally.

My personal gratitude goes to David Ward, SVP and Chief Architect at Cisco, for his commitment to my personal success and for instilling in me a self-confidence to be better.

To Damian Muzzio for guiding me in the logistics for this book project.

To our book reviewers we thank you for your prolific feedback.

Finally, a special thank-you to Mark de Jongh and the River Publishing team who shared a vision for what this book can be.

Special thanks to Thomas Cook, Christina Vargas, and D'lynne Plummer.

Endorsements

"The Internet of Women^{sm} provides genuine, serious examples from over 30 countries of how real life civic engagement, serious investment, and technology training gives women a voice and the means to get stuff done."

-Craig Newmark, Founder craigslist and craigconnects

"Technology companies need to constantly evolve to remain successful. At Vonage, this is one of our core values, to innovate relentlessly. To live up to that goal, we need to attract the best talent. It's well known that there are far fewer women than men in STEM professions, and to attract more women, we need to show them that there are incredible opportunities. One of the great things about this book is that it brings together the stories of many successful women and highlights not just their accomplishments, but the journey to success. I hope this book will inspire more women to choose careers in technology, which is incredibly important to ensure an ever-growing talent pool of technology experts."

-Alan Masarek, Vonage CEO

"In Saudi Arabia and much of the Arab world, women are outnumbering men as graduates in computer science and engineering. *The Internet of Women^{sm}* captures this incredible evolution in the region and highlights how technology can be an equalizer globally."

-HRH Princess Reema bint Bandar Al Saud, CEO of Alf Khair

"Our lack of knowledge about how our minds really work continues. We still say things like 'right brain' and 'left brain' when referring to either creatives or scientists, whereas technology was originally developed, and referred to by both. In fact, in the Victorian era, 'technology' was referred to as 'useful arts' this then evolved into the Greek term techne, 'science of art' and then came the German term, 'technik, 'translated ideas' which became technology. Perhaps in order to be an excellent technologist, one has to be a creative/artistic thinker. The more thought we give towards the emotional, artistic and creative sides to technology will surely result in more of a positive impact in long term versus just short term needs. *The Internet of Women^{sm}*, are just that ensemble. Looking beyond the horizon."

-Amy Redford, President of Boxspring Entertainment
-Clare Munn, CoFounder of Boxspring Entertainment

"For too long, women in the working world have been underpaid and undervalued. This is changing and it must if we are to produce economic growth that will propel our civilization forward in the 21st century. I am seeing increasing numbers of women in top management positions. Google is forcefully correcting compensation asymmetry as it establishes equal pay for equal work, particularly at the point of entry into the Google workforce. This trend should and must be reinforced – it is irrational to do otherwise."

-Vint Cerf, Vice President and Chief Internet Evangelist, Google

"We will soon enter an economy where tailoring jobs for people is more cost-effective than shaping people to fit job slots. From then on, raising the value of people will be the largest service market, larger than the entire world economy today. Because women are so undervalued, *The Internet of Women^{sm}* is well positioned to become a frontrunner of economic growth. With a high ROI on gender equality, holding back women from reaching their full potential will come at an ever higher price. The disruptive innovation economy is targeting gender inequality."

-David Nordfors, CEO IIIJ and co-chair i4j Innovation for Jobs

"We are at a point in history where women are leading the conversations around change from technology, social engagement, consumer products, media, medicine and every vertical you can think of. *The Internet of Women*^{sm} amplifies these female voices into a choir that can be heard from every continent around the globe."

-Joanne Wilson, Gotham Gal Ventures

"With the transition to a digital economy, the low representation of women among computer science professionals globally risks leaving not only women, but entire countries behind. *The Internet of Women*^{sm} shines a light on the increasingly pressing importance of fully engaging women in building and shaping our online experiences."

-Ann Mei Chang, Chief Innovation Officer and Executive Director, U.S. Global Development Lab, USAID

"The Internet is changing how we communicate and interact in deep and lasting ways. Technology, laws, social norms, and the substance of published materials all define how the Internet will shape society. It's vitally important that we embrace the opportunities the Internet offers, and simultaneously diminish the uglier sides of human interaction it can bring to the fore. The unique experience and perspective of women must play a central role in all aspects of how we build the Internet. If we – men and women – want to bring that about, we should pay close attention to what women have already achieved, what has worked well, and what women envision for the future. *The Internet of Women*^{sm} presents many compelling examples of women's leadership, and invites us all to continue in that exploration."

-Pete Forsyth, Founder, Wiki Strategies; Senior Editor, English language Wikipedia

"Empowering women is one of the most powerful engines of economic growth, and this book highlights the important role that technology can play in jumpstarting that engine."

-Adam Grant, Wharton professor and New York Times bestselling author of *ORIGINALS* and *GIVE AND TAKE*

"I was raised believing you had a moral obligation to make the world a better place, not just for yourself but for others. However, changing the world requires knowing the world, both its histories and the remarkable myriad of its present states. Change also requires patience, persistence and passion. And if you are lucky, you are not alone. *The Internet of Women^{sm}* reminds me powerfully that many of us are committed to making better worlds and that our stories and voices matter. The book is an important contribution to our conversations about the role of women in the world, about how technology might help create new possibilities and about the paths that are already being forged."

-Genevieve Bell, Intel Senior Fellow & Vice President

Foreword

The Internet of Women^{sm} is the next frontier, and it will be conquered. Women from all walks of life have demonstrated tremendous tenacity, strength and courage over the past decades, refusing to accept the status quo and breaking barriers in the face of stiff challenge. I've had the privilege of seeing this first hand, most especially as Board Chair since 2008 of Vital Voices Global Partnership. Over the last 15 years Vital Voices has searched out, trained, resourced, and networked over 15,000 women leaders across 144 countries. Now the leading global NGO for women's leadership, Vital Voices grew out of a U.S. State Department initiative following the historic 1995 United Nations Fourth World Conference on Women in Beijing. There, then-First Lady Hillary Clinton advocated for equal treatment and opportunity for women in every country. It was a pioneering step for women globally and a recognition of the work to come.

As the community of women's voices began to grow concurrent with advances in technology, the voices of women in science, technology, engineering, arts and mathematics (STE(A)M) grew as well. Creative women, digital women, and strategic women were finding ways to unlock the power of technology to solve some of the world's most pressing problems. Their stories are here in this book. Some of these women are highly educated and fortunate to be born in countries with great resources. Many of these women are not, but their lack of material wealth or resources is overridden by their determination to make this world a better place.

Here are stories of women like Akanksha Hazari, a recipient of Vital Voices' Economic Empowerment Global Leadership Award, who has created India's first loyalty platform for mass-market consumers and microbusinesses. Her journey is living proof of the power of possibility. Despite humble beginnings, Akanksha earned degrees at Princeton and Cambridge universities, returning home to India to ensure that her opportunity would bring value to others.

While working in rural India, Akanksha realized that even in areas where infrastructure and service delivery were inadequate, nearly everyone had access to a mobile phone. In 2011, she led a team of like-minded students to win the Hult Prize, the world's largest student competition for social good. She turned the $1 million prize into seed capital to develop the **m.Paani** business model, rewarding consumer spending at in-network partner stores with points redeemable for life-changing needs such as water, electricity, and sanitation.

Sarah Thontwa, born in the Democratic Republic of the Congo (DRC) and a survivor of civil war, is now a Senior Research Assistant at the International Food Policy Research Institute (IFPRI) and Co-Founder of Femmes Politiques (FEMPO). She works with women in the DRC to use blogging, crowdfunding, and ICT tools in politics (and beyond), to raise the cultural profile of women, and to fight marginalization. Technology has become a powerful tool as women struggle to transform communities and change national policies from which they are largely excluded.

In Guatemala, innovators like Karla Ruiz Cofiño are changing the way individuals and companies navigate, contribute, and participate in the world. Founder of Digital Awareness and MILKnCOOKIES, she has dedicated herself to creating a "collective intelligence" of global shared knowledge through the Internet to tear down obstacles to knowledge, and to empower women.

The Internet of Women^{sm}: Accelerating Culture Change provides many remarkable examples of women from all around the world who are creating change through technology. Each part examines an important subsection of women in technology: millennials, educators for the next generation, women in leadership, entrepreneurs, men supporting women, women at the forefront of new technological fields, and female pioneers.

It also reflects on how our institutions are changing. My home state is host to its own powerful example. The University of Wisconsin-Madison recently opened an Internet of Things (IoT) Lab, reaching across disciplines to expose women with non-engineering majors to technology, including holding boot camps for middle and high school-aged girls to show them that labs can be comfortable and fun. Research Director Sandra Bradley is a visionary and an activist, building confidence in high school and college-aged women that will surely change their path to success.

Leadership like this is the surest way to support and accelerate the digital knowledge of women. We have great leaders, but we need many more men and women involved in technology to join in this mission. This book proves that catalytic leadership generates powerful personal and cultural change.

We should be so grateful to the editors of this book: Nada Anid, Ph.D., the first female Dean at the New York Institute of Technology; Laurie Cantileno, one of the first female computer science majors at the New York Institute of Technology and now an executive at Cisco Systems, along with Monique Morrow who serves as CTO of New Frontiers and Engineering, and lastly Rahilla Zafar, a longtime writer and researcher focusing on women and technology in emerging economies including the Middle East, Pakistan and Afghanistan. Together, they bring a global perspective to the challenges of accelerating culture change and a determination that technology is the change agent for ensuring gender equality for present and future generations. Their vision and leadership is exemplary and very much needed.

Susan Ann Davis
Chairman, Susan Davis International
Board Chair, Vital Voices Global Partnership 2008–2016

PART I

Millennials Leading: Exploring Challenges and Opportunities Facing the Next Generation of Women in Technology

1
Building Communities through Technology

1.1 Introduction

Being born into a new era of technology gives the millennial generation an opportunity to redefine all the rules of the workplace. What the young women featured in this chapter have in common is that they were able to not only build borderless communities, but to see no limits on who can be empowered through the reach of technology.

1.2 How Shiza Shahid Is Supporting Female Founders and Their Critical Role in Reshaping the World

Shiza Shahid partnered with the equity fundraising platform AngelList to form the fund NOW Ventures that backs "mission-driven" start-ups with diverse founding teams including women. Shahid previously co-founded the Malala Fund.

What follows is an edited transcript of an interview conducted by Rahilla Zafar.

"It's convenient to paint the world in simple ways, such as by opposing good and evil, oppressed and liberated. But I think that's just not the way the world actually is."

I grew up in Pakistan, of humble origins. My father was born in a village and lost his own father when he was very young. He grew up quite poor. My mother grew up in the neighboring city of Faisalabad in a very religious, traditional family. She had an older brother but was herself the oldest of four sisters. In her community, women's education and careers were not considered or discussed beyond high school. When my mother graduated, she had an arranged marriage. She was told about my father in advance but met him for the first time on their wedding day.

Somehow, as my parents traveled and built their lives together, they saw the power of education. My parents really wanted my siblings and me to receive a great education. So they did something that was very unusual and very brave. My father was a naval officer, and while his benefits included relatively good and free schooling for us, my mother insisted on sending us to private schools. There was a time when my father was spending close to 50% of his income on our education. Sometimes he would even borrow money to pay our school expenses, and his friends would say, "You're playing outside your social class, and it's not a good idea." They would add that such schools were for families who could afford to send their children abroad for university.

Of course, these schools were also co-ed schools—schools where "boys and girls hang out and do things that are wrong." These issues and assumptions were a big deal in a place such as Pakistan, where "honor" is everything. However, my parents' controversial and significant decision changed the course of my life. We moved to Karachi and then Islamabad, where I attended a private school that cost 5,000 rupees (500 USD) a month, which was considerably more than the fees for other schools.

The Making of an Activist

As my parents' youngest child, I had more freedom than my older siblings and was very bored—in Pakistan I was like most children who didn't have extracurricular activities: you go to school and then come back home. But I felt that I needed more, so I began to follow various adults, parents of friends, who were involved in non-governmental organizations (NGOs) and who were doing cool things. Volunteering became a big part of my life. I was closely involved in earthquake relief efforts, for instance, as well as in organizing a march of 500 people in the judiciary protest against President Pervez Musharraf. I didn't have any tools, I was just taking action. I didn't consider myself an activist or entrepreneur. I had had no formal training. But that was my childhood, and I think being able to write about that passion and experience is what got me into Stanford University, with a full scholarship.

At Stanford I became restless. I was a young woman from Pakistan who was given a big opportunity, and I was there to change the world. But that's not necessarily the university experience. I was very passionate about bringing about change back home, but the culture in college felt more local and academic. So I kept seeking fulfillment in work back home. I'd go to Pakistan three times a year on holidays and build programs in order to create a social impact.

In 2010, the floods happened in the SWAT Valley in Pakistan, where Malala Yousafzai is from. I had met Malala and her family previously after seeing a news piece about her and had reached out, offering my help in their cause. That part of the country is very remote—and therefore not a priority for the government. I wanted people in the capital to mobilize; I wanted military and government intervention so that the schools could reopen. So I brought about 30 girls from SWAT, including Malala, with her dad's help, to Islamabad. I wanted these advocates for the schoolgirls in the SWAT Valley to be heard by people of influence. Thankfully, security eventually improved in SWAT, and Malala and other girls went back to school.

After I graduated from Stanford, I joined McKinsey & Company's Dubai team, and that gave me the opportunity to work and travel throughout the Middle East. Then in 2012, Malala was shot on her way home from school by militants opposed to her advocacy of female education. I had stayed in touch with her family and felt compelled to visit them. I cared about them, and they trusted me. When she was airlifted, I met her family in the United Kingdom (UK) where she was being treated.

Supporting the Creation of the Malala Fund

Malala's shooting became breaking news, and everyone around the world was horrified. It was a miracle that she survived, and everyone wanted to know her story. In the United States there's a narrative around the plight of South Asian or Middle-Eastern women that sees them primarily as victims. This takes away all of the nuances of who we are as women and how strong we are. Malala is not a victim, she's an advocate. I wanted to help her transition into her new life, now in the UK, and I wanted to help her tell her story in her own way. And that was the beginning: two books were published, and a film was made.

I said to Malala, "The world is listening. What do you want to ask the world to do?" Her father, Ziauddin Yousafzai, Malala, and I began the Malala Fund, which started as an effort to raise funds and direct them to issues that are important to the education of girls globally. I eventually agreed to leave McKinsey and build out something bigger with the Fund. It became two things over time: first, an advocacy organization that worked to persuade governments to ensure free, high-quality education for all girls from kindergarten all the way to the end of secondary school, and second, a funding vehicle for programs that supported girls in vulnerable parts of the world. In 2014, Malala was awarded the Nobel Peace Prize. It was both a happy and a sad time—Malala won the prize on December 10, 2014, and there was a horrific school shooting by militants in Pakistan just days later in which 132 children aged from 8 to 18 were killed. There was so much work left to be done.

Launching a Fund to Support Female-Led Innovation

I looked at my own trajectory. I was an activist through and through. I went into consulting after Stanford because that's where all the talent was going and was encouraged to go. But for some time it was the need for social change that was driving me. I realized that the most effective change is happening through social enterprises, and that so much more could happen if money flowed to them in the right way. There are outstanding examples around the world. One is Bridge International Academy, a female-founded chain of low-cost private schools. Another is Andela, which trains women and men in computer science in Kenya and Nigeria. A third, Acumen, which was founded by a woman, provides capital for start-ups. So what I focus on now, after the Malala Fund, is getting more investment capital to flow to early-stage enterprises that have been founded by women. I do a lot of speaking about, and research on, those models so we can put them out into the world and produce the most impact.

My goal is to enable social entrepreneurs to invest in and create ecosystems such as that in Silicon Valley, with a focus on servicing great ideas and solving problems. As part of this journey, I've taken the time to really understand where technology is at and spent time among tech start-ups. It's fashionable to use the word "technology" to scale education solutions, but what does that mean?

Technology has become synonymous with innovation, but it's a means to an end, not an end in itself. Additionally, impact organizations have to be sustainable. It's really about building something with a revenue model so that it can scale. And these organizations need investment capital. If we want to have an impact, we need to move billions of dollars, not tens of thousands. Real investment capital, as opposed to pure philanthropy, is what is required to make a big impact, and I am working to build a consensus to do just that.

Social entrepreneurship already offers a viable choice for education in many countries. In Pakistan, a high percentage of people go to low-cost private schools. We must acknowledge that the government is failing at education and that private schools are doing better. The government could issue vouchers to families so they can send their children to those private schools that offer high-quality education. That's an example of integration between private enterprise and entrepreneurs. Increased participation and better education lead to greater engagement on the part of women in the economy and in entrepreneurship. That will have profound implications on gross domestic product (GDP) and on culture.

Another example of social change from Pakistan: Sharmeem Obaid-Chinoy won her second Oscar for *A Girl in the River: the Price of Forgiveness*, a film about honor killings that was recognized by the government and has resulted in important changes to the law. That's an enormously positive step. In Pakistan, there are more women training in the medical professions than men. Now we have to ensure this translates into employment, and we have to leverage it—perhaps as a government marketing campaign—so that girls are allowed and encouraged to study and work.

It's convenient to paint the world in simple ways, such as by opposing good and evil, oppressed and liberated. But I think that's just not the way the world actually is. I think that in the future we'll look back at the way we now view ourselves, as countries divided by borders, as an archaic way of dividing a global civilization that is essentially interconnected in every way.

1.3 An Activist's Evolution into Banking and Technology

By Sarah Judd Welch, CEO of Loyal

Sarah Judd Welch is CEO/Head of Community Design of Loyal, a community agency that works with product, marketing and innovation teams to understand, engage, retain and collaborate with customers. Clients include General Electric, National Geographic, NYU and start-ups at all stages. In former lives, she worked for Hillary Clinton and a top-tier investment bank, and she believes that community is the future of economy and society on the Internet.

"To date, much of our technology has been built by white, upper-class men. This means that a significant portion of the world has not been considered in building the future."

Growing up in San Jose in the 90s meant being raised in a bubble defined by liberalism and technology. I didn't know any Republicans. My dad was a network-hardware engineer, my mom was in enterprise quality-assurance support, and all of my friends' parents worked in tech. My high school didn't mind when we skipped class to attend protests. I learned hypertext mark-up language—html—at a summer-camp program. I volunteered often. Gay rights, global trade negotiations, Carly Fiorina's leadership challenges at Hewlett-Packard, and immigration were all regular topics at the dinner table. My entire life seemed to be shaped by the belief that we could make the world be what we wanted.

As such, it was perfectly ordinary for me to spend the summer between middle school and high school riding the bus from the sprawling edges of San Jose towards downtown's worn-down, gang-ridden neighborhoods to spend my days volunteering as a tutor at an elementary school. I didn't think much about it—I enjoyed fumbling my way through Spanish and hanging out with the kids.

That's why it was a shock for my 13-year-old self when the dot-com bubble burst in 2000. It's true that this was more of a slow diffusion rather than the sudden pin prick of a balloon, though to me as a tween, it felt sudden, and of course, it was out of my control. Shortly after that summer, my mom drove our moving truck through that same neighborhood where I had volunteered. Like many families during that period, we were relocating to somewhere much cheaper. My mom pulled up into the driveway of a house I had never seen before. I hopped out of the passenger seat, and I turned around. There across the street was that same elementary school where I had volunteered. I distinctly remember that roaring sound in my ears and that dizzy feeling in my body—the feeling you get when the world suddenly changes. My identity shifted at that moment. I was now a part of the community I had once considered to be so geographically, economically and culturally distant from myself.

One by one, my friends' parents lost their jobs that year. Both of my parents were laid off. I was let go from my first summer job before I had even started. There just weren't enough families with money to enroll their kids in summer camp. The tech-fueled new-money mentality of "spend everything," combined with a failing economy, Enron-sized water bills, and a struggling State University system, meant that there was a huge contrast between my optimism about the world and the reality of that world.

Like many idealistic teenagers faced with vexing circumstances, I decided that I would do something about it. This wasn't an unusual choice in my family. We were all involved in the world. My cousins are all teachers. My grandmother was a veterans' nurse. My grandfather was heavily involved as a sponsor and counselor with Alcoholics Anonymous. My brother participated in the Seattle protests against the World Trade Organization (WTO), led marches around education issues in Sacramento, and participated in labor organizing. My mother turned an abusive marriage from before my brother and I were born into a life-long passion for teaching women's self-defense. For whatever reason, we all felt a duty to serve the world and make it better.

As a teenager, I didn't really know how I wanted to "serve the world," just that I would. I already knew at that point that I wanted to work for myself. I was ruthlessly independent in that young, arrogant way. I wanted to solve problems: homelessness, education access, economic opportunity... anything, really. And, I assumed that I could do it better than those before me. So, I left sunny San Jose and all of California's messy problems behind to study sociology and nonprofit management in New York City.

In Search of a Career Path

At Fordham University, making the transition to New York City was a bit rough at first. I wasn't as well-supported as my classmates and came from a very different social and economic background. For the first time, I met Republicans. I had friends who were pro-life. I shared a dorm with students from old money who had trust funds, and in stark contrast from the ethnic and cultural diversity of my recent world in San Jose, I was in a community that was primarily white and that assumed that I was just like them. Here, the assumptions about the world were different, and it was a culture shock for me.

For the second time, my world, my community, and my identity within it all changed dramatically. While I felt out of place, I learned over time that this was okay. There is nothing like feeling like the odd (wo)man out to push you to empathize with others, appreciate diversity and to turn your own biases, ideas, and beliefs inside out.

And boy did I feel naïve. Shortly after the start of my first semester, I realized that sociology *studied* problems, rather than solving them. I found myself constantly asking why and how. I wasn't satisfied with my classmates' explanations about why they thought being gay was wrong, or with the response from my high-school mentee that "college just isn't for me." I found myself digging deep into larger political, economic, and cultural problems that seemed too hard to solve on my own. I found myself turning towards politics, yet my interests didn't neatly fit into traditional academic programs.

Halfway into the second semester of my second year, I withdrew from the classes I would have taken the following fall. I was done. I was acing my classes, but my curiosity was unfulfilled. I felt like I was in an alternate, conservative universe on campus. I didn't see how staying would get me to where I wanted to go, even if I didn't really know what I was aiming towards.

At some point after that, I applied to NYU on a whim. My application essay was about a pair of shorts that didn't fit, and I wrote it the day on which the application was due. The year before, I had checked out an independent study program called Gallatin, and I thought that at the very least it would give me the freedom to study the things that I actually wanted. I got in, and without an alternative drop-out plan, I went to Gallatin.

The Making of a Banker

For the first time, I had peers who could also weave a conversation around health care, public policy, and economics in five minutes flat. I studied globalization, the philosophy of justice, racism, and political journalism. I read Plato, and then Adam Smith's *Theory of Moral Sentiments* and *The Wealth of Nations* back to back. I interned for Hillary Clinton's Senate '06 campaign and for the Taxi Workers Alliance. I helped candidates run for local office and was even elected to a local seat for the Democratic Party myself.

I found that capitalism or democracy themselves weren't the problem. They are just systems built by people with certain values, and all systems and all people are flawed. I learned to appreciate and to find curious others' values and beliefs rather than to judge. All systems incentivize and disincentivize specific behaviors and further the agenda of the values they are built upon. I felt as though I was really onto something: to change the world, you need to change the system.

Yet, as graduation approached, I was skeptical about the efficiency of the approach that I had been taking up to then to "changing the system." I started finding out how much my friends who graduated before me were making working in politics. Without the financial support of their parents, they couldn't make the rent. And, given that I was graduating with significant debt and without the benefit of wealth behind me, I knew that entering politics wasn't a viable option. Moreover, I was deeply disappointed in my experience in local electoral politics; I couldn't get anything done. I wondered whether there was an alternative path towards changing the system.

Around that same time, 2007, the market started to take another downturn. Even at that time, everyone was nervous. My classmates and I wondered whether we'd have jobs at all once we graduated the following year. By that fall, I had decided that I needed to explore other career options. I went to NYU's career center and walked them through my resume and what I had studied.

After remarking on my leadership experience and economics coursework, my career counselor suggested I talk to an investment bank that would be coming to campus the following week. I didn't even know what investment banking was. And that's literally how I got a job offer at one of the most desirable places to work in 2008 for ambitious undergraduates in a failing market. It was the only job offer I got, and I knew that I was lucky to get even that one, so I took it, accepting the offer on the last possible day before expiration. I cried about it: it wasn't the future I had imagined for myself.

I was miserable at the bank. I worked such long hours that my friends knew to call my office line directly instead of my cell phone. The Starbucks across the street from my office was on speed dial. The disconnect between my work and any kind of tangible meaning was depressing. Less than two months after I started, Bear Stearns filed for bankruptcy. Every day after that, I went to work thinking I was about to be fired. Two-thirds of my team was laid off.

Yet, despite myself, I fell in love with business at the investment bank. Maybe it sounds terrible, but I loved being a witness to the inside of the financial crisis, seeing the cause and effect of different levers of our economic systems as they were pulled up and down and the rippling political and social effects. It was the perfect case study in systems. And I loved the speed at which the market moved—much faster than the public sector.

Unsurprisingly, however, I didn't stay for long. But when I left, I knew that I would stay in the private sector. I knew that I wanted to work for myself, as I always thought I would. At the same time, I also knew that I didn't know anything about running a business. So, I went to my one friend who had started companies in the past and shadowed him for six months. At the time, my friend was working on an online platform for job matching which hit many of my sweet spots, including economic empowerment, and following my experience in investment banking, personal happiness. I was fascinated by the idea of online platforms actually making people's lives better. It wasn't through a social service and it wasn't through policy, but it was impactful just the same. Through technology, I could make people's lives better. Through the market, I could shape the world.

Merging My Skills with Technology

It was through this in-between phase of my career in 2010 that I became involved with both the tech community in New York City, and in hand, found my next career move at Catchafire, an online platform that matched professionals who wanted to give their skills to non-profits who needed their help, and later, TaskRabbit, an on-demand labor platform for chores and tasks.

Both platforms helped me to see that I didn't need to work in politics or on a large economic scale to impact people. I could make everyday people's lives better through platforms, both through the power and people-systems of the platforms themselves, and through the way in which the platforms did business. Good business is good for everyone.

This is where Loyal, a community agency that I founded in 2012, came from. Loyal works with brands to build and leverage communities online. I started Loyal because community is powerful. On the individual level, having relationships with others makes us happy. And on the macro level, communities create opportunities to create the world we want by working together. And even more so, on the business level, community-driven businesses are better businesses: they act as better stewards of their customers, and they tend to perform better, too. For example, check out Airbnb, Etsy and Lyft—all driven by community. In fact, according to the 2016 MIT study "What Creates Advantage in the 'Social Era'?", community is one of the primary catalysts of innovation, resulting in a 37% improvement in overall company performance.[1]

I look at Loyal's work in building communities as building value-based microsystems in which people interact together. These communities are mini-economies and mini-governments. And importantly, community thrives on diversity. Communities exist to connect people who feel that they are on the fringe, and to prove that there are others just like them. And beyond that, communities exist to expose us all to *new* ideas. Just as my exposure and participation in various communities in different stages of my life opened my eyes and shaped my view of the world, everyone can benefit from diversity in ideas. Everyone can benefit from community.

This is why I'm so excited about communities of women and increasing diversity in technology. To date, much of our technology has been built by white, upper-class men. This means that a significant portion of the world has not been considered in building the future. Yet we know that diversity in ideas that are developed via communities improves performance. Thus it's not surprising that companies with diverse employees perform better. In fact, women-led businesses perform three times better than the SAP 500.[2] And so it follows that increasing the diversity of the technology workforce will result in a stronger outcome for the future of the world—for everyone.

Community connects people to each other. Community thrives on diversity and generates new ideas. New ideas increase performance and result in better business outcomes. So if you want a better business and a better world, invest in your company's and your world's communities by bringing in more women. And, not just more women, but more diversity.

As it turns out, my career is not all that different after all to what I aspired to accomplish as an idealistic teen. The more connected the world becomes, the more opportunity we have to build community, and the stronger the world becomes. From bitcoin and Blockchain to freelance economies and shared knowledge bases, community is inherently democratizing and empowering. That's why it's so important for the Internet of Women[sm] to exist. With community, the Internet of Women[sm] can work together to create the world women want and deserve.

1.4 Young People Have Power: Millennials and STEM

By Karoline Evin McMullen

Karoline Evin McMullen specializes in connected technology and mobility ecosystems. She has been recognized as one of the world's most promising young activists by Dr. Jane Goodall, and received the President's Environmental Youth Award from President George W. Bush at the White House. She has also received the Gold Congressional Award from the United States Congress. She is a graduate of Yale College.

"As a girl in the 90s and early 2000s in the rural Midwest of the United States, I was relatively alone in my fascination with science, unsupported by the schools I attended and ridiculed by my peers for being more excited about chemistry kits than I was about boy bands or prime-time TV."

Science and technology are a portal to rapid progress—to the forefront of discovery, where new developments can have market-changing implications, and where a single disruptive invention can change the lives of millions. Although the landscape of opportunity for women in science, technology, engineering and mathematics (STEM) has drastically changed in recent years, and especially in the last 15, in the sciences there are still gaps between young women and young men in terms of levels of support, early education, and role models. Accordingly, I work every day to help bridge that gap in my community and around the world. I am inspired by the growth of strong female voices in tech circles both in the United States and abroad—especially in emerging markets. However, there is still a long way to go before we achieve gender equality in STEM.

My Upbringing

As a girl in the 90s and early 2000s in the rural Midwest of the United States, I was relatively alone in my fascination with science, unsupported by the schools I attended and ridiculed by my peers for being more excited about chemistry kits than I was about boy bands or prime-time TV. My parents were both fierce advocates of my interests, and taught me to be undeterred by the disapproval of others. Their unwavering support was likely rooted in their own educations and backgrounds. The son of an engineer, my father is still, 42 years later, the only National Merit scholar to have graduated from his high school in rural Ohio, and the only graduate of Harvard College and Harvard Law School from his town. My mother is a first-generation American and a Cornell University graduate. Her father was a basic-science researcher who had earned his Ph.D. and gone on to become a professor. After the Second World War, her family housed and supported other families in need, and even sent their children to college. Education was, and still is, the most valued and honored accomplishment in my family. Though I didn't recognize it as a child, one of the largest drivers in my life has been a devoted respect for the pursuit of knowledge.

In elementary school, I spent summers learning about programming and robotics while my friends tie-dyed t-shirts and played capture the flag at sleepaway camp. Most girls in fourth grade were cheerleaders for peewee football on Saturday mornings, but I preferred to spend that time going to the Great Lakes Science Center or reading. In my early teens, I recognized that science would allow me to pursue my passion while allowing me to make a substantial difference in my community. I launched an organization to educate the community about the local environment, through the lens of science. Through this initiative, I found boundless opportunities to explore things that excited me more than decades-old lessons in school. I began to teach others and make a difference, even though I was only a high-school student. Though I had to focus on my primary role as a high-school student, I was thrilled to have found new ways to learn outside the classroom, where I could apply useful new knowledge in ways that could actually affect the world around me. I felt strong, capable and respected through my work in the sciences. Every weekend, every night after class, and even sometimes early in the morning or during my lunch break, I thought of new ways to apply myself and affect my community.

I helped to create national curricula, through which I reached over 10 million students across the United States, teaching them about environmental, active citizenship and how science affects our everyday lives. I earned grants to fund the restoration of numerous environmentally devastated areas, plant thousands of trees, and reintroduce a healthy habitat that has since become a safe home to threatened and endangered species. I ran a fish hatchery and conducted original, funded ichthyology research. I helped develop a new exhibit at the city zoo featuring local fish species. I created park signage that has been installed along trails to help visitors learn while they explore. I wrote and published textbooks and curricula that were used in schools in three states. I was hooked on the impact of science, and was determined to broaden my scope of involvement. I started working with underserved urban schools to secure materials and teach hands-on science. I guest-lectured at universities and developed educational materials for elementary schoolgirls to introduce them to STEM careers in a fun and engaging manner. I felt unstoppable, because I was furthering local and national conversations rooted in science, even though many people had doubted that I would be able to or that it was a good use of my time, because I was "just a kid."

Although I attended a well-respected co-ed prep school, the teachers and administrators had actively discouraged and mocked me for my involvement outside the classroom, calling it a "distraction," and saying that a girl like me should just focus on her coursework and stop being so high-minded. Despite current trends, many young women still face this gender-based discrimination today around their interests in science and community impact. As a student, I was lucky enough to have the support of adults outside my school environment who saw the value of my efforts and who understood that, despite my youth, I was adept, driven and capable of accomplishing my goals. I firmly believe that this faith in the ability of young people, especially from their teachers, is essential to future diversity in STEM. Young people have power, and adults in positions to influence them would do well to recognize and support that fact. Doing so will increase the quality, legitimacy and potential impact of young leaders' work in the sciences, and their confidence in pursuing their interests without fear of failure.

My experiences in high school expanded my understanding of how science can bring positive change to communities and classrooms. It also inspired me to explore more ways in which I could become involved in science and technology once I enrolled at Yale as an undergraduate student. Throughout college, I continued to teach STEM curriculum to underserved students, and began to mentor young women who were interested in pursuing science and technology. I helped many young women see beyond their circumstances and earn scholarships to college, even when their families didn't believe that they could. I also became fascinated with computer science, returning to my love of technology and applied science. I conducted independent research in digital environments, exploring how technology affects social interactions and the free exchange of information. I learned how to code, explored the legal system that governs technical intellectual property, and I worked in several internships in technology both in the United States and in Europe. These internships made me realize that a global perspective would be essential to my future success and would help me approach challenges with more openness and creativity.

Entering the Workforce

After graduation, I moved to Los Angeles, where I lived for several years, working in the service of ideas that change culture. At first, despite my glowing performance reviews, my excellent relationships with clients and my consistently high-quality work, I dealt with disrespectful treatment based on my age, and occasionally, based on the fact that I was female. I faced being paid less and being afforded fewer professional opportunities than male colleagues who contributed significantly less to the team than I did. I decided that I had to be so good that no one could ignore me—that I had to work harder, be more prepared and think smarter than anyone else.

I learned to be thick-skinned and fearless and to take harsh, sometimes unfounded criticism as a challenge to be better than ever before. I became an invaluable resource to the team, because despite my youth, I had the most experience with science and technology. Within that role, I worked with clients from Harvard Medical School to the Large Hadron Collider to IBM Watson, helping to tell the stories of world-changing technologies and programs that will shape the future. I also spent many late nights and weekends working on behalf of those striving for a better, more equal future for women, including Princess Reema Bint Bandar al Saud of Saudi Arabia, Sheryl Sandberg, the ONE campaign, and the Open Society Foundations. I realized that I wanted to be the change agent affecting the development of these visions for the future, not just the spokesperson on their behalf, so I moved to New York to chase that dream.

I now work as a strategist, leading conversations about mobility, connected technologies and improving the lives of consumers. For the first time, my expertise and opinions supersede my age and gender. I am no longer viewed as just a young woman, but rather as a subject-matter authority whose opinions matter. I work with clients around the world, identifying opportunities for global partnerships and new technologies that will shape the way we live in the coming years. I am committed to creating a future defined by equality, environmental responsibility, democratic access to technology and social impact. Doing good in the world and doing well for shareholders are not mutually exclusive objectives. In fact, present market forces are increasingly incentivizing responsible business practices. It is now up to those of us with a platform and something to say to bring this future to fruition.

The most effective way to put this vision into action is through inclusive, diverse teamwork—teamwork that takes place in an environment where derogatory comments and jokes based on gender identity, race, ability and other unique traits are not tolerated. It is imperative that workplace leaders, team managers, educators, administrators, parents, troop leaders and others in positions of authority understand their power to set the tone for their communities and offices. Tolerance, respect and compassion can become the norm when they are demanded by those in positions of influence.

Some of the most powerful methods of cultural and economic transformation are consumer technologies, consumer media and the campaigns that surround them. Marketing, advertising, and media strongly influence the way we understand products, the way we interact with new technologies, and the way we see ourselves in the world. One shining example of responsible leadership in this field is my friend Lauren Barnett, who is the voice of some of the most influential women in modern American culture—the cast of Grey's Anatomy—and a contributing writer and editor for Barbie's Hello Barbie doll. As a Grey's Anatomy writer, she shapes strong female characters whose expertise in medical science is equal to that of their male counterparts, and whose empowered strength creates stability in the volatile environment in which the show is set. She has also been a contributing writer and editor, helping to maintain the voice of the modern Barbie character. As an interactive doll, Barbie interacts with young girls, providing a friend that the young girl can find relatable, and actively participating as a role-model character. Barbie's language has been specifically crafted around empowerment, making this Barbie the first interactive Mattel toy that helps girls understand that they can do anything—that they are independent, capable, bright and not constrained by gender norms; that they can excel in any discipline; that they can be at the top of the class; that they can be leaders, and that all of these possibilities are worthy of celebration. This direction has been maintained across the brand with a new Barbie campaign that shows girls as professors, veterinarians, businesswomen, archaeologists, scientists and coders. I even bought the first Barbie I've had since I was a kid when Computer Engineer Barbie hit the shelves. She still sits on my desk today, as a reminder to always embrace the joy that technology brought me as a child. This new era of Barbie is a huge step in the right direction, as its brand equity carries on as a massively influential force shaping the self-perception of young girls around the world. It is especially important, as the Barbie brand historically faced criticism around many problematic messages for young girls, from unhealthy body image to complaining about the difficulties of math.

Widespread messages lauding female capability in science and technology are increasingly present in everything from branded content to feature films. Though far from realising gender parity with men, these advances take the simple step of portraying women in positions and careers traditionally held by men, and that shows young audiences that their future has just as many options as anyone else's, regardless of their gender identity.

A significant contributor to female access to progressive thinking, technological education, and Internet entrepreneurship is the proliferation of ideas in the digital space. In this age of democratized media, content such as talks from TED and TEDx has become available to almost everyone, spreading thoughtful innovation from the world's thought leaders to mobile devices, computer screens, televisions and radios the world over. These talks have begun to create a common vocabulary of creative disruption, as, taken together, they have been viewed over a billion times. Open platforms dedicated to the free exchange of expertise, including Open Courseware, Khan Academy, massive open online courses (MOOCs), and Codecademy, allow anyone to acquire knowledge if they have an Internet connection and a device with a screen. Though this has long been a property of the Internet as a whole, the recent proliferation of dedicated education platforms has introduced a new way for anyone to learn, regardless of what resources they have, or of their geographical location, gender, or level of ability.

In order to build a more inclusive and diverse global technology community, we have to start with young women. They need to see their mothers, aunts, sisters, neighbors, favorite cartoon characters and heroes in leadership positions in every field—including STEM. They need to be encouraged and celebrated for their enthusiastic pursuit in every academic field—including STEM. Women of all ages need to be encouraged to believe in their abilities, just as I was encouraged to believe in myself in the face of doubt from others, and they will flourish. Today's female leaders need to take up the gauntlet and go out of their way to mentor and support younger women. Only through networks of collaboration can we all rise to meet the needs of the future. These goals, while abstract and impersonal, can become very tangible and substantive when we apply them in our own lives. And it is this active application—a living, breathing, daily dedication to making change—that will make our workplaces, our networks, our cities and our world better places, full of diverse perspectives.

1.5 Launching a Female-Inclusive Tech Sector in Gaza

By Iliana Montauk, Director of Gaza Sky Geeks, the first start-up accelerator in Gaza

Iliana Montauk served as the director of Gaza Sky Geeks (GSG) in 2014–2015. She graduated summa cum laude from Harvard University with a BA degree in History and Literature of France and the Middle East, and completed a Fulbright in Jordan focused on start-ups in the region. She has worked at Google, Monitor Group (now part of Deloitte Consulting), and Wamda (the TechCrunch of the Arab world).

Gaza Sky Geeks was founded in 2011 by Mercy Corps, a global humanitarian organization, with seed funding from Google's non-profit arm. A San Francisco Bay Area native and former Google employee herself, Montauk has experience that spans the tech sector and social-innovations fields.

"I called home in my first week in Gaza and told my parents, 'This place is more like Silicon Valley and Harvard than any other place I've been in my life.'"

Imagine a place 25 miles long by 5 miles wide with a beautiful coast and great Internet infrastructure. Add to it a highly educated, urban population, known for its leadership, work ethic, optimism, and willingness to take risks. These people have a deep love of technology and food, and a desire to be global leaders in innovation, business, and social impact. Sounds like Silicon Valley, doesn't it?

This is Gaza.

A Language Nerd Joins the Tech Sector

I arrived in Gaza in November 2013. Gaza Sky Geeks, an innovative program within the global organization Mercy Corps, had asked me to come manage the launch of its start-up accelerator—the first in Gaza.

If you had told me when I started studying Arabic in high school in 2000, that I would eventually be running a start-up accelerator in Gaza, I would have laughed with disbelief, for two reasons.

First, I was a language and literature nerd, not a person aiming to work in the tech sector. Back then, I thought that going into tech or start-ups was only for hardcore computer geeks such as my brother—the type that wore Costco jeans and shoes and kept their long hair in an unkempt ponytail. I was more likely to become a journalist, academic, or doctor—or so I thought.

Second, even though I was so eager to learn about the Arab world that I was taking Arabic classes before September 11 (when Arabic-language programs barely existed in the United States), I had a negative conception of Palestine. I assumed Palestinians were violent people who threw rocks. I planned to travel all over the Arab world except Palestine.

Both perceptions changed with time. I joined Google after college in 2006. By then, the tech sector seemed like a welcoming place that attracted top talent and solved world problems. In 2013, when Mercy Corps recruited me, I was working at the equivalent of TechCrunch in the Arab world, seeking a way to harness the tech sector, create a positive social impact, and make use of my languages. By then I had also met many Palestinians, and of course, there is no better way to let go of negative stereotypes than to meet the people they are about. I learned quickly that Palestinians are a hardworking, educated, and enterprising people—in fact, Palestinians are known throughout the Arab world for being leading businesspeople, the ones who built the oil sector in the Gulf.

I called home my first week in Gaza and told my parents, "This place is more like Silicon Valley and Harvard than any other place I've been in my life."

Google Visits Gaza—and Sees Its Potential

Gaza Sky Geeks had been launched by Mercy Corps in 2011 with seed funding from Google.org, precisely because others from Silicon Valley had had a similar experience. Gisel Kordestani, Mary Grove, and other "Googlers" had been making their way around the world to conduct developer outreach. After several attempts in 2008 and 2009, they were able to enter Gaza (whose borders are almost entirely closed) thanks to Mercy Corps, which had access to Gaza through its humanitarian mandate.

They were blown away. Later, they would still say that Palestinians had struck them as some of the most entrepreneurial, enthusiastic, resilient and adaptable yet isolated people in the world.

Thus, Google.org donated nearly $1M to launch what was then called the Arab Developer Network Initiative. Its agenda was to stimulate tech and entrepreneurship in Gaza. By late 2013, when I joined, some of the start-ups that had been through our competitions were receiving seed investment offers from regional funds.

An Isolated Start-up Community

It's December 2015. In two days, our start-ups will pitch at a competition in Jordan. This event is one of our top priorities for the year: run by a well-known Silicon Valley accelerator and one of our strongest partners, 500 Start-ups, it will draw both regional and international visibility. In addition, this is the first time this batch of start-ups will travel abroad.

Back in November 2013, my team arranged for Gazans to travel to Jordan so our start-ups could meet with their potential investors. We had never previously taken our start-ups out of Gaza. I was called into an emergency meeting with the Israeli Defense Forces because all of the "Gaza exit permits" we had applied for had been turned down. I learned then that Gazans are not allowed to leave Gaza via the border with Israel, since Gaza and Israel are at an official state of war, unless they meet one of a select list of exceptions: they work for a humanitarian organization, they are ill and have permission to receive treatment in Israel, and so on.

Since the border with Egypt is almost always closed and the border with Israel is open only to a select few, most of our start-up founders have never left Gaza before. This is their biggest challenge as founders: not power outages, not ongoing conflict, but rather, extreme isolation.

My team and I asked the Israeli Defense Forces (IDF) in 2013 to add Gaza Sky Geeks to their list of exceptions. After an internal review, they did, and our start-up Geeks started traveling to Jordan via Israel regularly, and from Jordan onwards to other parts of the Middle East, Europe, and so on. That border crossing became their lifeline.

At the 500 Start-ups competition, one of our start-up Geeks received the second-place prize. His radiant face appears all over Twitter with pictures of a $5,000 check, Dave McClure, and a regional representative for Microsoft, the sponsor of the event. The winner was the same start-up founder who was pulled aside for a 12-hour IDF security interview a week earlier. "A start-up pitch must be easy in comparison to that security interview!", I think, smiling as I see his overjoyed expression.

During that same trip, our start-up founders went to a movie theater. For most of them, it was the first time in their lives. Founders also expressed surprise at the fact that Jordan has 24/7 electricity—a basic piece of knowledge they should all have, since Jordan constitutes a target or test market for most of them. Even Gazans who follow start-up trends online are clearly affected by the physical isolation. When I sent my Gazan start-up accelerator manager to Europe for the first time, in 2015, he came back telling me "big data is a new trend that is all the rage now!". This was the top start-up leader in Gaza, a man who regularly reads *TechCrunch* and other publications, and he was hearing about big data several years after it had been a buzzword in the United States.

Gaza reminds me of my mother growing up in Poland, behind the Iron Curtain. She could not travel to the West because of her country's politics, not her own.

When she received an orange or a banana, she would treasure it. When she would meet foreigners, she would soak in the experience, learning everything she could about what was happening elsewhere.

Similarly, when I brought the first copy of *The Lean Start-up* to Gaza, my team members tenderly held the books as if they were precious. And the best way to accelerate their companies, I've learned, is to keep them as closely connected as possible to the outside world. Although access to the Internet in Gaza is quite open (unlike some parts of the world, such as China), knowledge simply does not transfer online as fluidly as we assume. Even in tech, the human-to-human connection is key.

How to Give Birth to a Tech Sector in a Frontier Market

The strategy of Gaza Sky Geeks has pivoted several times since our launch in mid-2011. For the first two years, we ran *ad hoc* tech training and activities to inspire potential entrepreneurs in a part of the world where the concept of launching a start-up was not widespread or encouraged. In 2014, we added a co-working space and a start-up acceleration program that included brokering seed investments for our start-ups with a set of investment partners. In 2015, we expanded our investor network, ran our first ever incubation program (a prerequisite for entering our acceleration program), and began exploring a second line of business, thus creating immediate employment opportunities for tech talent in Gaza via distance working, outsourcing, or freelancing (our pilot was with Microsoft).

The key lesson during this evolution is that, while the quality of our local talent is high by regional standards, there is nevertheless a large gap between what students learn in college and what they will need to run a start-up successfully. Gaza's unemployment rate has skyrocketed since its borders closed. In 1999, it was 20%. In 2015, it was around 40% for the general population, and as high as 60% for the youth.[3] This means that most potential start-up founders have never had any work experience. When I arrived in Gaza to launch our start-up accelerator in 2013, there were four start-up incubators already present, but as far as I could tell, nobody in Gaza had heard of lean start-up, minimum viable product, product management, or user acquisition.

Thus, we put our start-up teams through an intense period of knowledge transfer and experience before presenting them to our investment partners. During our "incubation," start-up teams went from idea stage to launching a minimum viable product (MVP).

This development in our strategy is exemplified by our cornerstone annual event. Start-up Weekend is a three-day competition during which teams give birth to a start-up idea, and ideally, launch it. In the first year, we expected the winning team to depart with a viable start-up and thus awarded them $20,000. That is the size of a seed investment in the region, with the equivalent purchasing power of approximately $500,000 in Silicon Valley. At present, we view Start-up Weekend as our outreach event. It is an opportunity to draw talent into our pipeline and influence others in Gaza to be working on start-ups. We give the winning team approximately $500 because we recognize that teams at that stage need a large investment of knowledge capital before they can make good use of financial capital.

The challenge is that, in order to invest knowledge capital effectively, you need to have the right staff and strategy. The key word here is ***effectively***. There are lots of humanitarian organizations and local universities offering training in places like Gaza. But to invest knowledge capital in a start-up team that has the potential to become a successful business, you need a staff that understands investments and the private sector. Only such staff can make effective decisions about whom to select for the knowledge-capital investment (i.e., the incubation program) and how to transfer that knowledge capital (i.e., what to teach and how). While we are still developing our own program, our traction so far has demonstrated that, with the right talent and strategy, the knowledge gap can be bridged sufficiently in a relatively short amount of time (6 to 24 months).

When I arrived at Gaza Sky Geeks, my manager tasked me with developing a business model for this work. Eventually I concluded what one of my best mentors would later confirm: building a start-up pipeline in an emerging market is not a for-profit business. We have developed several revenue streams and aim to recoup 20% of our costs. For instance, our investment partners are for-profit entities that invest in start-up teams directly and then pay us for our services in accelerating those start-ups. Nevertheless, the cost of inspiring and preparing potential start-up founders in Gaza will outweigh the revenue we are able to generate. As for equity, it is a numbers game that also takes time: we would need hundreds of start-up investments annually and a decade in order to reliably make a profit. Finally, a key component in our model of impact is influence, which does not necessarily generate revenue. For example, one sign we are doing our work well is that other incubators in Gaza have adopted our curriculum.

Thus, planting the seeds of a tech sector in a new environment requires patient (philanthropic) capital. This is the education business, and just like Silicon Valley started with government grants decades before it created exponential job growth, entities such as Gaza Sky Geeks will need to exist for a decade or more before Gaza becomes the Silicon Strip. If we do not begin now, we will lose a whole generation of talent in environments such as Gaza. This will be a loss not only for their communities, but for ours as well—Silicon Valley needs their talent just as much as the Middle East does.

Though our business model depends on philanthropic funding, it is still important to select our donors carefully—and here, the global tech, entrepreneurship, and private sectors are important partners. Our current funders include Google for Entrepreneurs, Microsoft, Techstars Foundation, Skoll Foundation, PalTel Group, Bank of Palestine, and Bayt.com. These private sector sources provide "smart money": they measure our success according to metrics that meaningfully guide our work, share curated and targeted networks of fantastic potential partners, and advise us by sharing insider tips or observations they have gathered through their deep experience. Their engagement expedites the success of tech entrepreneurship in frontier markets. In addition to those private-sector partners, our model relies heavily on partnerships with for-profit investors. We broker investments between them and our start-ups. Any investor can choose to make an offer to any start-up. I have learned that the potential for financial investment has to be there in order not only to motivate entrepreneurship, but also to ensure that our activities are rigorous and that they produce commercially viable start-ups. We shaped our incubation program at Gaza Sky Geeks after our first year of running an accelerator, because we then understood what our investors would be seeking. Without that experience, it is too easy for a "start-up incubator" in an emerging market to be launching glorified university-graduation projects.

How We Nurtured Women Start-up Founders

Mariam Abulteiwi was a fourth-year computer-engineering student at a top university in Gaza, and she had never heard of start-ups. She attended our Start-up Weekend in 2011 because she had heard Google was going to be there. When she returned to our next competition in 2012, she came prepared to pitch an idea. It was essentially an Uber before Uber was well-known. She kept working on the idea until she became the first female CEO in Gaza to close an investment for a start-up.

Mariam had never planned to become the CEO of a business. Neither had Abeer al-Shaer. Abeer was studying computer engineering and English literature in 2015. Because she was ambitious, she was completing two separate bachelor's degrees simultaneously, and planned to become a university lecturer or a teacher—the professions in which she had seen women. After completing the Gaza Sky Geeks incubation program, she said, "I now realize I was meant to be a CEO, or at least a manager. I love building things. If my start-up does not succeed, I'll get an MBA. By launching a company, I feel like I'm shaping society, not just observing it."

When I arrived at Gaza Sky Geeks, approximately 25% of our participants—predominantly in our outreach activities—were women. This is high for the tech sector globally, but Google for Entrepreneurs challenged us to increase that even further.

Through focus groups, we learned that women had no stigma with regards to science, technology, engineering, or math (STEM) fields. Many women told me, "In the United States you think men are better in math and science? That's ridiculous. Most of the top students in those fields were girls when I was graduating from high school." We realized we had an advantage over some parts of the world and wanted to harness this opportunity to create one of the most inclusive tech sectors. We also felt a responsibility to do so since our tech sector was just being born, a blank slate. If we could successfully build it inclusively from the very beginning, we would not have the harder work of undoing bias later on.

Women were facing some barriers, however. They said they did not have the knowledge they needed to launch a start-up, which we interpreted as a lack of confidence since we saw that men with the exact same level of knowledge did not hesitate. This we have been addressing through regularly scheduled activities specifically targeting women. For example, when we ran an introduction to an Internet of Things (IoT) event for the general community, only a couple women attended amongst dozens of men. But when we ran an IoT event just for women, 40 attended eagerly and asked very intelligent questions. In addition to growing their knowledge, these events build a strong, supportive community of women who boost each other's confidence later in co-ed settings.

Women also said their families would not support their endeavors if they drained the family's meager income. In Gaza, $400 may be a family's total budget for the month, and an individual coming daily to our co-working hub can easily spend $50 a month on transit. Men are more often able to receive this funding, either because they make the financial decisions in their families or because families are more eager to fund their career endeavors even if the payoff is not immediately clear. We began giving women small stipends ($50 to $100 a month) to participate in our activities and found that their family's buy-in immediately increased. Women told me personally that this is what enabled them to participate or prevented them from dropping out.

We were also conscious of women's needs in many other ways: scheduling events at times they could more easily attend (most women in Gaza are expected to be home by dark), prominently featuring women in our outreach materials, and weighting our acceptance to ensure 50% participation by women in our main early-pipeline activity (Start-up Weekend). As a result of these measures, our overall women's engagement rate is now around 40%, which I believe is an impressive statistic globally. (We still have challenges securing investments for women-led teams—cracking that one is one of our top goals.)

1.6 Going beyond Gender and Tackling "Diversity Debt" in Technology

By Andrea Barrica, venture partner and entrepreneur-in-residence at 500 Start-ups, one of the world's most active, global, and diverse micro-venture-capital firms (micro-VCs)

"I've come to the conclusion that misogyny in tech is not just about women—it's partly about femininity."

My career in technology was an accident. For 10 years, I dreamt of becoming a linguist in the CIA. The summer after I graduated with a degree in linguistics from UC Berkeley, I planned to move to China. One day, out of the blue, I got a call from my freshman roommate from my first college, Simon's Rock, a natural-born entrepreneur and newly graduated computer scientist, Jessica Mah:

"Andrea, we just finished YCombinator, and I am raising money right now to build a financial dashboard. I need your help. Come to Mountain View, and help me build this company. I need a right-hand woman."

Less than a week later, I moved to Mountain View and began leading operations, customer happiness, marketing, and all things non-technical. To be honest, I didn't have any interest in tech at the time and had never heard of YCombinator. I was more interested in supporting my friend. I was perplexed why Jess would choose me over her MBA friends, but she was sure she wanted someone "who lacked the institutional imperative"—advice she had read from Warren Buffett.

The company we founded, inDinero, began with a bang: it raised more than $1.1 million. It grew fast, and it generated a lot of hype. Then, we realized the original "Mint.com for business" freemium SaaS model was not going to work, and had to fire everyone (including customers) and start over. I stayed with Jess and our other co-founder, Andy, and built the new model from scratch with them. We had to bootstrap after spending most of our funding. Then, we changed the dying model: instead of just a self-service dashboard for small businesses, we developed a full-service platform that would replace the bookkeeper and accountant, and relieve a major back-office pain for growing companies. Jess studied for the IRS EA exam to learn about filing taxes, and I did the bookkeeping, accounting, and payroll for the first group of companies, with the help of a few CPAs. My next role was all about sales—something else I had never done. I managed to generate a million dollars of revenue in about nine months, mostly through referrals. I helped build inDinero with Jess for over three years before leaving to pursue other dreams, though I definitely found my heart in entrepreneurship. Today, inDinero is almost six years old and has almost 200 employees worldwide. It's like having a child, it becomes a teenager, scales, and gets a life and a personality of its own. It's been gratifying to watch.

Mentoring Other Founders

After inDinero, I started pitch coaching to pay the bills. Over the last few years, I've coached thousands of entrepreneurs from all over the world, from Silicon Valley, to the American Midwest, to Brazil, Saudi Arabia, Ghana, Turkey, and more. As part of my commitment to help more women in tech, I go out of my way to coach women and minorities.

All founders are different. However, in general, when I work with female founders, we work extra hard to boost confidence, cut out apologetic language, and work through personal insecurities of whatever kind. Many women over-explain their credibility, as if they're trying to earn their spot on stage or in an investor's office. People from emerging markets or non-confrontational cultures share this, too. Men also need this coaching, but much less so in my experience.

I love my role selecting and helping companies at 500 Start-ups. Not only is it a fun place to work with truly awesome and smart people—one of 500's core values is "Be diverse. Be diversified." Internally, our investing partners are over 30% women, and half of our staff are women. Over one-third of our portfolio comprises of companies with at least one female founder, and we've invested in over 150 female CEOs. I also love that not many people know this about us. This hasn't been a public-relations play: it's just who we are. We've had so many female founders and CEOs in the community—for example, in Batch 14, Havenly, a female-led company and a marketplace that connects interior-design professionals to people who want to redesign or decorate their homes. Globally, women are economic powerhouses and control big markets, so it just makes sense to invest in women and markets targeted toward them.

I've come to the conclusion that misogyny in tech is not just about women—it's partly about femininity. Creating a more inclusive tech community should include conversations about valuing the other spectrum of skills—high empathy, vulnerability, emotional intelligence, communication skills, storytelling—not just checking boxes by hiring more diverse people in the company. Again, diversity is not just about more female or minority founders. We have to start valuing the things women and people of all ages, religions, backgrounds, and so on bring to the table. It's a cultural issue, which is why I started talking about "**diversity debt**."

Diversity debt is a term I have been using to describe the challenges start-ups face when it comes to building a diverse team, and more importantly, an inclusive culture in start-ups. Just as engineers can accrue "technical debt" when they push out sloppy code, companies can also accrue diversity debt over their lifecycle. The more people your company hires until you have a diverse team (meaning an array of genders, LGBT, socio-economic backgrounds, ethnicities, ages, levels of able-bodiedness, and so on)—the more diversity debt it has accrued.

Recommendations

- **Start early**. People often ask me: when in a start-up's life does it make sense to prioritize diversity? From personal experience, I would say that

1.6 Going beyond Gender and Tackling "Diversity Debt" in Technology

debt starts to accrue around the fourth hire, speeds up around #10, and becomes REALLY HARD after #20. If you don't believe a homogeneous team is beneficial for the future of your company, start early—even when it doesn't feel urgent. Homogeneity becomes harder to change as your company grows. Small cultural and process changes can make big differences over time.

- **Build an inclusive culture from Day 1.** Define cultural fit, and be specific about what exactly your core values are and what message you want to send to current and new team members. For example, does "work hard, play hard" as a company value manifest itself as 14-hour workdays and wild weekend drinking adventures as a team? If so, it's also a huge repellent for anyone who is NOT a young, extroverted 20-something without kids or any desire for balance.
- **Examine your job postings** for language that alienates women, minorities, parents, or older people. This includes highly exclusive language and aggressive language. "Hire More Women in Tech" is a fantastic resource and primer.
- **Revamp your interview process.** Beware of whiteboard technical interviews or alcohol-based social test outings with prospective employees.
- **Publicly offer and describe benefits** on your website, and include domestic-partner benefits, maternity, paternity, and adoption leave—even if no one needs it.
- **Try to interview at least one diverse candidate for every major role** you're hiring for. It's a version of the Rooney Rule strategy that helped the NFL increase coaching staff diversity. The key is to take them through the full process—and it doesn't count if you rule them out before meeting them. Why?
    a) You are giving the candidate a better chance to be fully vetted.
    b) You become accustomed to interviewing candidates with backgrounds different from yours.
  Facebook has started doing this for a select set of roles.
- **Watch the office housework!** Currently, women do most of the office housework. Pay attention to who takes out the trash, orders food, and stays on to clean up after events. Create a rotating schedule for these tasks.
- **Educate yourself** and your team about unconscious bias.

1.7 From MIT to Supporting Women and Internet Penetration in Jamaica

By Ayanna Samuels

Ayanna holds three degrees from the Massachusetts Institute of Technology (MIT). She has broken many barriers in Jamaica, where she has run her own business since returning from the United States. She is the only female rocket scientist she knows on the island, and is deeply committed to the socio-economic development of the Caribbean.

I am a descendant of enslaved Africans. In fact, slavery was abolished only two generations before my birth. My story as a young Jamaican of African descent in a post-colonial, post-slavery society is truly atypical. Many Jamaicans have a strong Eurocentric definition of beauty, a legacy of our enslaved past. Bleaching of the skin is en vogue, and it is quite common to chemically alter one's hair to make it look like white people's hair. It is often said on the island that 21 families rule Jamaica, and that hardly any of these families are black. Some context is necessary to help explain this phenomenon, given that approximately 90% of Jamaicans are of African descent. When slavery was abolished, Afro-Jamaicans had zero land ownership, 100% unemployment, and no education (a tool for empowerment that had been strictly forbidden during slavery). White slave owners and their families, along with other immigrant groups, secured much of Jamaica's property. They continue to hold a tight grip on the country's resources, and hence its wealth, up to today.

Given all of this, outside the safe cocoon of my home, the messages I received when I was younger reinforced the message that I was worth less than those of my compatriots who were lighter-skinned than I. In fact, there is still a popular notion that *anything too black is not good*. Still, my family always taught us that who we are is enough, and as such I never sought to change any of my physical attributes to become more like European-like. I was therefore not one of the popular girls in primary or secondary school. However, my parents remained sources of strength for me, as they always gave me strong messages that reinforced my self-worth.

Throughout Jamaica, unemployment among women is very high. This has negative effects, chief among them being a strong dependence on men. That belittles women and makes them vulnerable and exploitable. It also reinforces negative messages about gender roles to the children these women raise, often

1.7 From MIT to Supporting Women and Internet Penetration in Jamaica 31

in single-parent households. Many women are taught, however, that they are chiefly defined by their sexuality, and it is this that will allow them to have children with a better-off man and thus create a better life for themselves. But having to feed those children only increases their dependence on men and creates a vicious cycle, in which they are just as vulnerable and exploitable as ever. Growing up in this social dynamic, I was committed to changing this paradigm. I realized that one way to do this was to get a great education.

All my life I had dreamed of being an astronaut, and early in my high-school career I set my sights on MIT. It was a daring dream: I knew of only two other people who had been admitted. Neither was black, and neither was a woman. Still, I was fiercely determined to be admitted—I simply knew I had to give it a try. And, much to my delight and amazement, I got in.

When I arrived at MIT, I was not convinced there was a seat for me at the table. An aunt who had gone to Harvard warned me that some people might view my admission as a fluke of affirmative action, the result of some "quota." "They're wrong," she told me. "You earned your place. Ensure you always sit at the table and make it clear that you belong there." Had she not prepared me, my experience at MIT could have weakened my spirit. The first time I doubted myself was at MIT: I often felt unaccepted. As the only black woman in my class, I experienced moments with professors and other students that made me feel like I simply wasn't in the room. Once, a member of the janitorial staff thanked me for picking up a trash bag the day before, assuming I was the new janitor. When I explained I was a student, she thought I was a new employee that was now hallucinating about being a student and looked at me with a sense of pity for the future of my career as a janitor. It is experiences such as these that speak volumes about the work that is left to be done and the inequity in the STEM field for descendants of Africans, especially women.

But the reality of all this inequity really hit home when I graduated from the Aerospace Engineering Degree Program in 2005. I was told that I was the first black woman to earn a master's in science from the Aerospace Engineering Program since 1972—and I am not even American. That motivated me to dedicate my time, efforts and passions to lifting the tide for all ships in Jamaica.

I actually have three degrees from MIT: a bachelor's in Aerospace Engineering with Information Technology, the aforementioned master's in Aerospace Engineering, and a further master's in Technology Policy. I was also a finalist for a Rhodes Scholarship.

My family worked hard for MIT to be an option for me. It almost doesn't make sense that I was able to go to MIT, considering that my maternal grandfather made a living from picking up bottles, while my grandmother sold fruit at a market. My father came from a lower-middle-class family, the son of a pastor and seamstress. When he was ready to go to university, the government made post-secondary education free. That's the only reason he was able to go to law school and become an attorney. My mother worked with him as a paralegal.

During my time at MIT, I learned that astronauts generally had a particular area of expertise that they would bring to a mission. I then began to search for what my area of expertise would be and came upon satellite communication technology, and saw the potential it had to bridge the digital divide for emerging economies. I was hooked. I saw how this technology could help developing economies leapfrog over barriers to socio-economic development, and thus reduce poverty. I was determined to use this technology in a symbiotic relationship with other information and communication technology (ICT) to contribute to the socio-economic development of the Caribbean. There are more poor women than poor men, and women are more heavily impacted by the digital divide. I thus felt that women in the Caribbean should be able to start taking advantage of the lucrative opportunities that ICT presented.

Upon returning to Jamaica in 2005, I made a number of key observations. There is generally a large socio-economic gap between men and women. Even in a relationship where the woman is working, men tend to make the decisions: as I noted earlier, many women have been raised to feel that what they bring to a relationship is their sexuality, while men must take care of them and be in charge. Thus, even in a relationship where the woman may be financially well off in her own right, she may still feel incomplete without a male partner, or if she does not have children. That kind of thinking can also lead to an acceptance of abusiveness on the part of men, and even to the notion that such abusiveness is a sure sign of love.

Gross gender-based inequities are also in evidence, but in a different way, in the workplace, including at senior levels. Although there are quite a few women in middle management, men predominate on Boards of Directors and in top decision-making roles. In Jamaica, it's a man's world, with men calling the shots.

Partly as a result of these inequities at home and at work, a lot of Jamaican women who could be game changers choose to leave the island instead, thus exacerbating the already significant brain drain.

But that's not the path I decided to take.

Giving Back to Jamaica

I decided to move back to Jamaica after MIT because I wanted to contribute to the meaningful and sustainable development of my island. This is where the anchors of happiness are. It's a culture I know well and love.

There is so much potential to develop Jamaica. When I'm 90 and sitting in my rocking chair at home, I want to be able to look back on my life and know that I devoted it to that development as much as possible. I thus founded a consulting practice focused on using ICT and technology policy for socio-economic development and the eradication of poverty. I also conducted research on gender equality in ICT, and have given many a lecture on this topic as I am keenly interested in the intersection of gender and ICT/STEM policy.

Examples of Life-Altering Applications of ICT: Fighting Health Disparities with Mobile Technology

Let me give one example of the potential of ICT in Jamaica: technology has the potential to connect women with much-needed health resources. Specialized health apps for women, for example, could be channels for providing the latest information about reproductive and sexual health. I see a lot of potential, for example, in an "ovulation" app, which would help women to track their ovulation cycles, and thus further help ensure that every child is a wanted child.

There are also opportunities for telehealth services to make medical help much more accessible: mobile platforms could make doctors and nurses available by phone 24 hours a day to offer affordable and personalized medical advice. Participants could pay by the minute with credit loaded onto their mobile phones. In Jamaica, many people are already set up with prepaid mobile phone service. Imagine the possibilities for women being able to privately discuss sensitive health issues with a certified healthcare practitioner at a fraction of the costs they now have to pay, and without the need to travel.

Another opportunity here is facilitating the use of ICTs by the Caribbean diaspora to help them manage the health of relatives resident in the Caribbean. For example, non-communicable diseases are the leading cause of death in Jamaica, consequently, the diaspora is keen on providing assistive drugs to their relatives for diseases such as diabetes or hypertension. Applications which would allow for the purchase of drugs with mobile credit loaded solely for the purpose of a specific drug to specific person, would help reassure the diaspora that monies sent for medication for their loved ones will be used for that purpose.

In Jamaica, mobile penetration stood at 110% in 2011 and is likely way beyond that now. However, the high cost of mobile broadband Internet data access continues to be a problem. Two leaders in the affordable mobile space are Huawei Technologies and BLU Products. And then there are local providers who sell self-branded phones. Some of these brands have scaled relatively well but have a less-than-stellar reputation when it comes to how well they actually work.

Not surprisingly, advertising plays a major role in the extent of scale and this has meant some low-cost phone providers still have not enjoyed the levels of scaling one would expect given their price. At the low end, smart phones can cost anywhere from USD 30 to 65. However, this is still too high for many people, especially women, many of whom are digital natives (the generation of people born during or after the rise of digital technologies). Cellphones would have to cost between USD 10 and 15 in order for everyone to be able to afford them.

But the cost of cell phones is only part of the problem: even the cost of mobile voice calls is quite steep, and the extent of digital inclusion could also be increased by making Internet data access less expensive. As I noted above, the high cost of broadband is still a major issue.

The tablet community, for example, has been infused with so many options that are cheaper. I want to believe that even in the next two years, we will see a continued reduction in the prices of smart phones, mobile voice calls, and broadband Internet. Digital inclusion is, without doubt, needed if all Sustainable Development Goals are to be met.

It is crucial that governments in the Caribbean be strong enough to say to foreign direct investors, or FDIs, "I know you're bringing in tax money, but you must show a commitment to ushering in the benefits you afford (digital inclusion in this case) across the full spectrum of social classes." In low-income communities, access could be provided via schools, post offices, and community centers as a part of the FDIs' corporate social responsibility initiatives.

A far-reaching application, which does not depend on data access and that has proven to be a game changer in other countries is mobile banking. The percentage of people in Jamaica with little or no access to banking services is very high: 34% have no access at all, while 52% have limited access[4] (source: Dawn Elliott, 2011). There are thus a considerable and growing number of opportunities to take advantage of the intersection between mobile telephony and digital currency (electronic payment systems). There are barriers to the broad availability on the market of mobile banking, and the first step is to get a solid understanding of these. Research conducted by Dawn Elliott in 2011[4] found that the barriers to the adoption of mobile banking are a function of literacy, fees, trust, relevance, financial standing and prohibitive due-diligence requirements. Thus, without interventions to address these barriers at the institutional, educational, socio-cultural, national-policy and regulatory levels, mobile banking will never realize its full potential in the Jamaican market. Financial regulators also have strong concerns about money-laundering. Safeguards are, however, available, and following best practices is one way to implement these.

Studies have shown that both mobile telephony and the use of electronic payments instead of cash lead to a rise in GDP (critical for a country as indebted as Jamaica). Indeed, a number of studies concluded that a 10% rise in mobile subscribers in emerging markets will lead to a 0.6% to 1.2% increase in GDP in those markets, because of the productivity gains associated with communication as well as new jobs.[5] (Source: International Telecommunications Union, ITU), "Speech by ITU Secretary General", April 6, 2010, and Indian Council for Research on International Economic Relations (ICRIER). *Now is absolutely the time* for Jamaica to fully embrace mobile banking.

Integrating Technology into the Curriculum

Tablets are available in some Jamaican schools, and the One Laptop per Child (OLPC) program has been implemented on the island. The program should help foster talent starting in primary school. Jamaica's innovation ecosystem is quite active with the island often winning international technology competitions. The key thing is to make access to technology and mobile applications more accessible to those in lower socio-economic brackets. Working with schools is a great way to achieve this.

All this said, incorporating technology into education has only just begun in Jamaica. There has thus been little time to gather lessons learned. Sister countries Antigua and Barbuda have done a very good job at this, and provide sterling examples. They have been able to interweave technology into the effective implementation of their school curriculum. To do this, they made a close connection with their short-term goals. The key has been pre-planning and communicating to stakeholders how technology can link them to what they already care about. As a result, Antigua and Barbuda have been able to meet the goal that most students share: finishing strongly at school.

Promoting Gender Equality in STEM

Unfortunately, women are still in the minority in computer-science classes. In Jamaica, primarily at the university level, these subjects are still viewed as the preserve of male students. There has not been much pro-active work in tackling gender inequity in STEM fields.

We need to raise more awareness of gender stereotyping so we can tackle that, too. Efforts to redress gender inequities in STEM fields should start even in pre-K and include national gender-awareness drives.

Another sister island, St. Lucia, provides an instructive example of how to get things done on this front: It has actively sought the views of key stakeholders to involve them in gender-equity efforts. They know we need to do more gender mainstreaming at the level of both policies and their implementation.

Jamaica's national ICT policy does not mention gender. This gender-blind policy must be mainstreamed to address the inequity problem and better serve the country's future development.

I'm very excited about the 17 United Nation's Sustainable Development Goals: On September 25, 2015, Member States adopted these goals to end poverty, protect the planet, and ensure prosperity for all as part of a new sustainable-development agenda. I'm particularly pleased to see those goals that entail tackling inequality, including gender inequality.[6] I have been involved in many conversations on gender inequities in ICT. For women, we're talking not about sewing machines but about increasing their capacity to take advantage of the engine for economic growth that is the ICT industry. The gender-parity movement really speaks to the responsibility we all have—whether we're working in government, in a multilateral organization or the private sector, or are simply individual agents of change—to promote gender equity in the Caribbean so both genders can make strong contributions and see that they each have a seat at the table. A critical action item at this point is to develop national gender-equity policies, and see to their implementation across the Caribbean.

Efforts to achieve gender parity must be all inclusive, regardless of one's socio-economic status. Throughout the region, holistic education is needed for re-socialization regarding gender stereotypes. Given observed needs, education should focus on reproductive health, growing e-commerce initiatives, facilitating the growth of innovative indigenous and cultural projects and of various other forms of art and craft, encouraging the growth of mobile apps to solve local problems, and indeed fostering a paradigm shift so individuals can see that they can build thriving businesses in any area where they can meet one or another market need. Particular focus should be on educating women from middle- and lower-middle-class backgrounds, who make up the majority of women in Jamaica and the Caribbean as a whole.

This educational drive must also respect cultural roots. This is especially important because it can happen that those who are more educated reach out to those who are less well-educated in a way that says, "It's my way or no way at all." However, people don't want their way of life to be disrupted and often don't want to be relocated from the place they call home.

The OLPC program can help spark change and create entrepreneurs who develop homegrown solutions. One does not need an MBA in order to be able to identify a problem and develop a solution right at home with the tools that technology provides. The best advice I can offer from my experience is this: we're all here on earth for a reason, and we can all be effective agents for change. To my fellow Caribbean women, I would say that what determines your worth is not whether you are from a rich or a poor country, or the color of your skin. Never allow others to let you feel inferior for any reason, and certainly not because you're from a "developing country."

1.8 Gender Balance and Technology in the UAE

By Amal AlMutawa

A catalyst, third culture kid and a social butterfly

Amal is the Chief Happiness & Positivity Officer at the Prime Minister's Office of the United Arab Emirates. Besides the exciting happy projects, Amal also works on many initiatives including The World Government Summit and The World Drone Prix. She is also an alumni Global Shaper from the Dubai Hub (World Economic Forum initiative).

Her passion for Youth Empowerment led her to get involved in several impactful mentorship programs in the UAE. She volunteered in many different initiatives including being a coach for High School students as part of the Hamdan Bin Mohamed Students Personal Development Programme and a mentor with the local initiative, e7—Banat AlEmarat. She also volunteered with INJAZ (part of Junior Achievements) where she mentored students who founded a company which won best startup and best CEO in 2012. She was also a care-taker for an elderly woman lady part of waleef program, an initiative under the Community Development Authority of Dubai.

Amal comes from a technology background where she first worked as a Network & Security Engineer, managing IT related Projects at eHosting DataFort (part of Dubai Holding)—being the only female engineer at the time. She then spent a year working with du, a telecom operator in UAE, as a Technology Security Expert. She subsequently moved into marketing communications where she headed the eCommunication Department in the Government Communication Office under the UAE Ministry of Cabinet Affairs. Continuing on in Government, she moved into Innovation and Special Projects.

Amal completed the Sheikh Mohammed Bin Rashid Young Leader's Program and holds a Bachelor degree of Science in Computer Science from the American University of Sharjah, UAE.

Working in the Public Sector

I currently work at the Prime Minister's office of the United Arab Emirates. The UAE has gone a long way when it comes to recruiting women in the public sector. It's empowering to know that we have eight female ministers in our government at the moment. The first female minister is Her Excellency Sheikha Lubna AlQasimi who was appointed as Minister of Economic and Planning in 2004 and now holds the position of Minister of State for Tolerance. A few other cabinet positions held by females are the Minister of State for Youth-who happens to be only 22 years old, Minister of State for International Cooperation, and Minister of State for Happiness. At our office, three out of four of the Deputy Director Generals are females. Our Director General is also a female. The office is 75% to 80% females.

The transition of women being heavily involved in the work place is quite interesting. Just like you see in the past in the west, it used to be boys club. Men would meet up after work to continue to discuss business matters. The same applies in UAE. In the old days, men would meet at the *majles*—a place of gathering for men- and that's where discussions around business and politics took place. What I noticed with the transition is that, in the old days, it was a boys club, like the Majles. It was taboo for a woman to be there. But now you see more and more women in a business environment attending such *majleses*. It's important to note that in respect to our culture there are still the men only gatherings at the *majles* for social events. I remember when Sheikha Lubna became the first female minister here—that was a big deal culturally. Her presence at the *majles* at Cabinet meetings was judged by some people as it was against the norm. Lots of people were very proud and many young women, me being one of them, took her as one of their models. Now we have more and more women taking similar roles. The number of women on boards is also growing. So the perspective on and of women is changing exponentially.

In government, we have a lot of females in the workforce including a female fighter pilot serving in the Air Force. Women in the UAE are in leadership positions, they work hard, and have proven themselves countless times. I've seen many incredible women in the government who have impressed me with their action and vision. For them it's more than just about proving themselves—they want to make a positive change and a difference in this world. It's incredible.

There is a museum for women in the old part of Dubai that displays the history of women in UAE and all their achievements. It has records of women being active in UAE dating back to the 1950's capturing their role in real estate, land sales, humanitarian, and social roles. The museum also features art exhibitions by women. This reminds me of a project I was part of that gave me the had the privilege of meeting the first Emirati female photographer—she's incredible. Steve McCurry did a project for *The Empty Quarters*—a photography gallery—called the Seven Princesses, and I was selected to be part of it. The project involved him taking photos of around 20 Emirati women who are accomplished which included the first female photographer. She's in her mid-80s now and in a wheelchair. She learned how to use a camera before she went to the first grade as an adult her first encounter with photography was in the 1950s. She started going to school in the 1970s when she realized she needed to help her own children with their homework. She's won many awards and I'm very proud to have someone like her in our community who has a beautiful story to tell.

My Background

My story begins with the number of schools I attended which is 10 in total due to the nature of my father's job as a diplomat. It taught me a lot including how to be adaptive to continuous change and extremely social since I had to make new friends almost on a yearly basis. In the UAE, I was between two of the emirates until the 3rd grade. Then we lived in Syria for six years. After my dad's service in Syria, we moved to Morocco in 1995. I graduated from high school in 1999 from the Rabat American School and that's when I moved back to the UAE. My dream was to move to the United States and get a bachelor's degree from MIT but my parents weren't "cool" enough at the time though. Especially since I'm the eldest daughter. However, with time, my parents' perspective on living abroad changed and my sisters got to go study and work abroad. My youngest sister is currently working for Price Waterhouse Cooper in London. My other sister got her masters from the UK and lived for about 2 years in the Netherlands as part of her job at Shell. I'm looking to go get my masters in Europe in the near future.

When I moved back to UAE in 1999 to continue my studies, the best school at the time and I believe still is in the region is the American University of Sharjah. I ended up with Computer Science as my major. It was new program with fewer than 30 students. I actually didn't want to study computer science. I wanted to study arts or architecture but my dad was not convinced with my choices. I remember him telling me, "who's going to hire a lady architect? Or what would an artist do for a living?" Now I'm glad that I didn't do either one of those major since I'm extremely happy and blessed with the knowledge that I got from engineering and how to it shaped me today. Given my bubbly, carefree personality, the choice I made really helped me with my career growth and to shape the way I think when it comes to problem solving. So when I had to make a choice at the time of what major to pick, I wondered, "what's new and exciting?" It was computer science at that point. So I thought to myself, "let me do that—I'm good at math. I managed to build my first website in school using html and it was fun." I was always at the top of my class in math in all the schools I've attended. I also scored 6 on the IB HL exam along with only two other students in my school (the highest score is a 7 which no one got at my school). My love of math also led me to take all the math courses offered at the university and I remember asking for more advance classes.

After graduation, I worked as a network and security engineer at an IT Service Provider Company for five years. They were the best years of my life when it comes to career and I continuously shocked and impressed a lot of our clients. As a woman in my region, female IT professionals was rare and very few did the labor work of being an engineer. I used to go to clients with a Cisco router in one hand and a switch or a firewall on the other. I've had clients asking me if I really knew how to mount them on a rack and if I actually know how to configure those devices. At the company we had about 40–50

1.8 Gender Balance and Technology in the UAE 41

people. I was the first female engineer they hired and besides the PA of the CEO, there wasn't any other girls in the company.

My boss at that time didn't want me to join his team given that I'm a girl. The story of that is quite interesting. I got interviewed by the CEO of the company who later told me that the one of the reasons he hired me is because I played varsity basketball. He said I've never met an Emirati woman who plays basketball. My line-manager however, was forced to accept me since I was hired by the CEO and I heard later over some gossip that he wanted someone "who worked hard, not a girl." He used to give me a lot of hard work just to make me quit. But I impressed him—every time he gave me something to do, I just went and did it and I kept asking for more challenges. He was sincerely one of the best managers I've ever reported to. He continuously pushed his team for growth. I loved the team I worked with. I even remember how we had Sudoku competitions and I won a couple. Also, as I became a more senior engineer, I used to compete with one of my colleagues on solving network problems for clients. The entire team was great to work given that they were from all over the world which was kind of home to me. We had people from UAE, Sudan, Jordan, India, Pakistan, Australia, Canada, and Germany. At that time, I was the only woman engineer, but by my third year with them, they started hiring more women in marketing and then later some engineers. My role at the company shifted to the project management side as it catered more to my needs of being a people's person. I loved to translate business requirements into IT words for the engineers to understand. I was the perfect bridge between technology and business since I was able to speak both languages.

After that, I moved to DU where I worked as a security expert. DU is one of the two ISP's in the UAE. My job consisted of being the ambassador of security across the different departments of DU. It was an exciting time as DU had just been announced, and they were growing exponentially at that time. There were a million and one projects and we needed to make sure that all security requirements were adhered by. It was still very male dominated but the number of women was growing with some impressive engineers as well.

Later in my career I moved to work for the public sector which at the time was a way of settling for me since I had to quit my job in late 2007 for some personal reasons and then rejoin the workforce early in 2009. You can imagine how tough it was at the time to find a job right after the financial crisis. I wanted to continue to work for private sector. Again, some unplanned events lead you to the most exciting times of your life. I've had the pleasure of working on extremely exciting projects in the government. I've had a front row seat to watch my government transition to a better version of itself year after year.

When I started working for the public sector, I started off as head of e-communication, and at the time, I was managing a couple of websites, including his Highness's website as Prime Minister of the UAE. We worked on one of his first e-sessions on the website with both the media and public. It was beautiful to see his Highness get involved with the public. He had announced that he would be answering the questions of the media and public through his website. The amount of questions that came through was incredible.

Social media was getting picked-up by some governments at the time. I worked with the team to develop a social media strategy for the federal government. We encouraged a couple of public figures to go on social media. Part of my job role was to managed the government communication network. It was an online portal for the communication teams across the federal government. The portal extended to a quarterly meetings with the communication teams to exchange knowledge and get to know each other amongst some trainings as well.

In 2012, the buzz word was no longer social media, it was innovation. Given my background in technology along with my passion for new ideas and people empowerment, it was the perfect transition for me. I moved to the innovation department with the sole goal to make UAE government more innovative. Our role consisted of two elements: how to make the employees more innovative internally and, how to make the UAE government more innovative. One of the project I led was to document innovation cases across the federal and local government.

The next transition for me was to move to special projects, with the focus on how to make our office a happier environment along with other projects. I became a trainer/facilitator on delivering happiness in the workplace. I've conducted a couple of workshops; it was a beautiful transition to use my different skills from my engineering days and apply it to human behavior. Not long after that, I've been appointed as the Chief Happiness and Positivity Officer as part of the governments direction with Happiness.

Exciting Initiatives from the UAE Government

In 2017, visitors of the UAE will be able to step into the future through the Museum of the Future. We're launching the Museum of the Future. It will be a permanent home for the world's greatest innovations, and will work to stimulate and incubate imaginative solutions to the challenges of future

1.8 Gender Balance and Technology in the UAE

cities. It will bring together the brightest researchers, designers, inventors and financiers under one roof. This will give women an opportunity to work in technology with touch points to daily practices. How to humanize technology maybe?

We have several awards that have been running for the last three years: Drones for Good, Artificial Intelligence for Good, and Robotics for Good. There continues to be amazing submissions. For example, a group of women from a local university came up with a drone that dissipates fog build up on airport runways. Another award we have is the mGov award. This award with many categories aims to make public services digital and easy to use from your mobile phone. The award, like it's latter, is not just for the UAE, it's international and regional.

The UAE is the only Arab country that is going to space. The project is called "Alamal" meaning Hope. The plan is to send unmanned probe to Mars by 2021.They've got a couple of female engineers up in there as well.

It's awesome, you can see a female presence in everything that we do here that's technical. It's not a man's world anymore.

Working in the Private Sector

Few Emiratis join the private sector. Instead, most take government jobs. The government pays a lot more—sometimes double or triple the salary—than you would get working for the private sector. You also get shorter working hours when you work for the government-on most days. So why would someone choose the private sector over government? I pushed both my sisters to work for a private companies because you get a different type of experience. For example, in my current government job, our office is super dynamic and we get to work on exciting new projects. We do a lot in a short span of time.

However, other government entities in the country may not offer the same kind of fast paced and interesting experience. Also, most of the government entities are mainly Emirati people which is not a bad thing but if you want an international environment, a government job may not be the place for that. The private sector in the UAE in most cases, offer a very diverse environment given that we are a melting pot for the region in a way. Even on an international scale, lots of Europeans, Americans, and Australians call the UAE home. The private sector, however has an issue with culture barriers when it comes to Emirati girls. Some young women are extremely shy and attended all girls schools all their lives so for them to integrate into the workforce, sometimes needs a little more attention and more time than others. The culture within a private company may not always respect the fact that girls may need a little privacy. For example, something as simple as shaking hands with men can be a challenge. A lot of local women don't like to do it.

Some Muslim women from different countries have the same challenge. In the private sector they will learn it's accepted, but may not feel comfortable with it at all. And, depending on which kind of company they work for (for example, European companies), the teams would go for drinks right after work. The social gatherings for an office often revolve around drinking. For a lot of local women, that's not something they're comfortable with, nor do they want to be seen in that type of place. It creates barriers from that perspective. Nonetheless, I'm seeing a huge shift for a lot of women who are now working in the private sector.

Some of the private companies, although not all, look at locals as just a "number" that they need to have to fill a required government quota. They don't really invest in teaching them anything because, to them, locals are a bit more spoiled. I've personally met some local women that *are* somewhat spoiled; all they think about is, 'I need a fancy title, I need a fancy office.' Instead what they need to do is to work hard. There is a small segment that have a sense of entitlement, but at the same time, I've seen super high achievers who are amazingly impressive people. I've seen both sides of the coin.

Future Goals

My future goals include opening up a school that teaches kids self-empowerment. Nowadays, you can Google anything, information is out there. But how to utilize that information and develop soft-skills—that's what's needed. Public speaking, negotiation skills, how to interact with people, how to communicate, how to listen to people, how to do data analysis, all these soft-powers that are becoming more, and more important each day. I can see some schools starting to invest in that already which is great. The school I'd like to open will only cater for teaching soft skills and something more than a traditional science class. I would like to enable students to be anything they desire. Making them creative, investing in what they are truly good at. For example, if someone is good at communicating and public speaking, they would be cultivated to help shape a future in the PR field perhaps.

Recommendations

Involve women and see what will happens. It's a simple as that. It should not be about quotas. It's not about numbers, it's about quality of people that you bring on board. It's about passion, it's about leadership, being visionary, about education. In UAE, women tend to be more educated than men, and they pursue higher education a lot more than men, so why not utilize them. 95% of girls and 80% of boys who complete their secondary education enroll in a higher education institution in the UAE or travel abroad to study.

1.9 Millennials Moving the Needle on Gender Equality

By Noa Gafni, CEO and Founder, Impact Squared

Noa Gafni is the Founder and CEO of Impact Squared, a consultancy that empowers communities to increase their impact. She has over a decade of experience working with leading organizations, from the World Economic Forum to the United Nations Foundation and Coalition for Inclusive Capitalism on their digital strategy, community activation and impact evaluation. Noa holds a BA from Dartmouth College and an MBA from London Business School. She is a Fellow of Social Innovation at the University of Cambridge.

My parents met in the computer unit of the Israeli Army. As I was growing up, there was always a computer at home, and I remember playing around with an early version of the Internet. It was called Prodigy, and at the time, it was hard to understand why it was different from the rest of my computer games. As much as I rebelled in middle and high school (purposely not taking a series of classes on computer programming), the sense of familiarity with technology stayed with me.

When I was in college, the Internet started to shift from a place where we could find information to a place where we could connect with others. As a psychology student, I was fascinated. I spent a semester at New York University as a research assistant to Professor Katelyn McKenna, one of the founders of Internet Psychology, and I wrote my senior thesis on the ways in which teenagers interact online. It was 2005, still in the early days of social media, and that term didn't even exist yet.

Inspired by the findings in my thesis and the online tools I saw that were connecting people in meaningful ways, I founded a social network for women, which combined an online platform with real-life events. After college, my friends and I dispersed, and found ourselves, for the first time, without a structure to meet new people. Many of us were in organizations that were primarily male-dominated and weren't finding work to be a comfortable space in which to connect with others. It was hard to make friends and meet like-minded people. We helped women meet friends-of-friends, connecting both online and in real life.

It never became "Lean In", but I learned a lot about networks and what makes people engage. I then went to work at Hearst Magazines as the head of social media. I launched online communities for *Cosmopolitan, Marie Claire*, and *Country Living*, as well as a cohesive strategy for the rest of the Hearst Magazine portfolio. It was amazing to see how established brands—some of which had been around for more than a century—understood that a huge shift was happening and that they needed to get a new generation of women on board. The most successful brands, such as *Cosmopolitan*, were the ones that combined an online community with mentions across their website and magazine. They understood that networks exist as a mindset, not a channel.

After completing my MBA, I joined the World Economic Forum as the Head of Communications and part of the founding team of the Global Shapers Community, a network for exceptional millennial leaders. The World Economic Forum was founded on a multi-stakeholder approach, bringing influential voices from all sectors to the table. As half of the world population is under the age of 30,[8] the World Economic Forum realized that youth are a key stakeholder—a group that will be most affected by the decisions made by global leaders.

Shapers are connected locally as part of a hub, a group of local leaders who meet in person on a regular basis. Select members of the community also meet at World Economic Forum events. But the Community as a whole never meets in person. It's a combination of digital touchpoints, a clear vision and informal connections that take place both on and offline, that makes the Community such a success. By the time I left, we were based in over 250 cities in 130 countries worldwide. We're now based in almost 500 cities around the world. It's amazing to see how a network can be so strong across geographic boundaries and language barriers, and despite involving so few structured interactions.

It's also unique in that women make up 50% of the Global Shapers Community at World Economic Forum events. Women make up fewer than 20% of participants at Davos[9] (the World Economic Forum's Annual Meeting), but the Global Shapers Community is the *only* group that's achieved gender parity. And Global Shapers are using this existing advantage to put forward ideas to the larger World Economic Forum community. At the World Economic

Forum's meeting on Africa recently, Global Shapers championed an initiative called "Internet4all" to close the digital divide (those who are not yet online). The digital divide affects women more than men, and access could be a game-changer.

As millennials take a bigger role in global conversations, women have a greater opportunity to participate in the conversation. They don't have to break the glass ceiling to be offered a seat at the table; they have to be leaders in their sector early on in their careers. Hopefully, these opportunities will enable them to close the gender gap later on.

I currently run a consultancy that advises organizations on how to create and nurture networks for social impact. One of my biggest clients has been +SocialGood, an initiative of many well-known foundations, including the United Nations Foundation and the Gates Foundation, that unites change-makers around the power of innovation. My role is to create a sense of community among globally connected influencers, to help share ideas of what's working in global development, and to empower them to adapt it to their local communities.

One of our Connectors, Ruba Al-Zubi (based in Jordan) is using the Sustainable Development Goals framework to increase the number of women engineers throughout the region. Michelle, the founder of ImpaQto and a +SocialGood Connector based in Quito, is trying to create an ecosystem of social entrepreneurs. For her, the +SocialGood community is a tool that connects local innovators with one another as a part of large global conversations.

Michelle believes that breakthrough innovation will come from unexpected places. She points to microfinance, which was founded in Bangladesh, not a global financial hub, and to the open-source mapping tool Ushahidi, which was founded in Kenya. According to Michelle, innovation happens when innovators are close to the issues of their local communities.

Women, who have often been considered as recipients, not creators, of innovation, are key to making this shift happen. And connecting them through networks makes this shift happen faster, as leaders such as Michelle and Ruba share ideas with each other and implement them in their own communities.

It's also exciting to see what happens when these conversations take place online: you don't need to meet someone to be inspired by them. As we look at getting more women involved in STEM, we need to think about what women are already doing online. Women are more frequent and more engaged users of social media. How can we meet people where they are and have conversations on those platforms? How can we empower young women who are already interested in getting into the technology world and help them become ambassadors? I love working with clients to build connections and create these important conversations.

I'm driven by this work because it's constantly changing: it forces me to learn more about technology, human dynamics, and how the two connect. Technology is only as interesting as the applications that we, as humans, find for it. I'm excited to see how emerging technologics, from artificial intelligence to blockchain tracking, will help us leverage technology in ways that make our lives better and more equitable, and give more people the opportunities that I've been privileged to have.

1.10 Supporting Women in Embracing Technology in Kurdistan

By Banu Ibrahim Ali

Banu Ibrahim Ali is an Information Technology graduate from the American University of Iraq-Sulaimani (AUIS). She is currently working at a private company as IT-Specialist in Slemani. Banu is an alumna of the Iraqi Young Leaders Exchange Program (IYLEP-2009) in the United States, the International Youth Forum Seliger 2010 in Russia, the National Model United Nations (NMUN-2013) in New York. She is an AMENDS Fellow (American Middle Eastern Dialogue at Stanford), Co-Organizer of Startup Weekend Slemani (SWS), and a former Co-Chair of the AUIS Alumni Association. Banu is also a Community Leader at LeanIn.Org.

I remember how, around 1999, computers first arrived in Kurdistan and everybody was talking about them. I was 10 years old at that time. Two years later, my brother and I decided to sign up for a computer-training program we had seen advertised. We were the youngest students in the course. Since then I have been hooked and have loved technology. I knew from that moment that I would never want to pursue a career in medicine or law or any other discipline traditionally valued in Iraq. My only dream was to pursue computer engineering.

When I graduated from high school in 2008, the American University of Iraq-Sulaimani (AUIS) was one year old. I heard that they offered information technology (IT) as a major, so I applied and was accepted, and started at AUIS that fall. People asked me what I planned to study, and when I said IT,

1.10 Supporting Women in Embracing Technology in Kurdistan 49

many had no idea what that was. Technology and IT were developing here in Kurdistan at that time. During Saddam Hussein's regime there was no access to the Internet, computers, satellites, or international TV channels. Those who owned televisions had access to local state-controlled channels only.

When I began my program, there were only two other women in the class. I had friends telling me that I shouldn't study IT because it's hard and not very suitable for women. They encouraged me to consider other alternatives. But two years later, I saw that more women were joining the program. I felt so happy because when they joined, we began teaming up to do class projects and develop technologies together. For instance, we developed Windows apps, Android apps, and websites. During my second year at AUIS, I tutored new students, especially women, because the percentage of women with an interest in IT was low. In 2010, one of my professors asked whether I wanted to volunteer as a teacher in an enrichment class about the basic mark-up languages (HTML and CSS) to new students who were interested. I taught this class once a week after school. I was excited to be doing this because the students were eager to learn and full of great ideas. With the right resources, they would be able to create innovative technologies.

Keeping this in mind, my friends and I organized the first Start-up Weekend in Slemani in January 2013. It was the first event of its kind to be organized in the region for entrepreneurs to team up, build their ideas, and present them to a panel. Start-up Weekend was an event for the community: whoever signed up could join, no matter their religious backgrounds or ethnicity (including Kurds, Arabs, Shia and Sunni Muslims, and Christians). At the same time, some of the participants teamed up with people that they were meeting for the first time. Attendees responded well to this event. I was not expecting a large number of women, but we had more than 50 participants, and I was happy to see many women participate. In fact, an all-women team ended up being one of the top three finalists. Their start-up idea was called "Back to Tradition" to encourage the use of traditional and local businesses.

A day before the event, we had an SWS trainer and judge from Google who came to teach some of the registered participants HTML5, CSS3, and JavaScript tips and tricks. It was a great hands-on training for developers on the teams. This trainer, Ibrahim Bokhrouss, was from Morocco and had worked at Google's Zurich office since 2007. One of the activities in his workshop was a design thinking challenge to come up with a nice Android animation themed with a specific Google product. The activity result was paper prototypes. And Ibrahim suggested that teams or individuals try to implement their prototype in code. One participant who followed up and coded the prototype was Rawand Nasih Fatih, a computer-science graduate. He then ended up getting an internship from Google the following year and joined Google as a full-time employee a year later in the Switzerland office.

We organized the Start-up Weekend three times and during the second and third events, we had more female participants. The ideas they had were great and although they were not all related only to technology, technology was always incorporated in them. For instance, Our Land was an idea to provide a space for females so they can express themselves more freely, Child Care Center was an idea for children to stay occupied while their parents are busy, and Antiqueium produced handmade and machine-made souvenirs.

Based on the success of the event in Slemani, Start-up Weekend has been organized in five more cities in Iraq, and is ongoing. And other, similar initiatives have emerged. At present, the enthusiasm for entrepreneurship has really taken hold, and this is very exciting to see. The good ideas that have come from Start-up Weekends around Iraq include the animated TV show that reflects the lives of Iraqis and aims to help improve communities, EasyNet, an application that helps you get the best Internet provider around you without having to try them all. Another one is Kurdistan Hikers, a hiking society that offers weekly hiking tours all year round in the Kurdistan region.

I graduated from AUIS in December 2012 and, before I started my first job, I traveled to New York City for the National Model United Nations. The first time I heard about Sheryl Sandberg's book *Lean In* was when it was launched in March 2013. I was in New York and I bought it. I did not know much about Sheryl Sandberg at that time, but she became my role model. After reading the book and learning more about her, I started the first Lean In Circle at the American University of Iraq-Sulaimani (AUIS) with the help of Kelly Peeler, the CEO of the start-up NextGenVest. Kelly is a Harvard graduate and worked with the Kaufman Institute to develop programs in entrepreneurship in Iraq. She also introduced me to the Lean In team.

Our circle began meeting monthly to discuss a given topic and share our stories with each other. What made me very proud is that we had men joining the circle, moderating it, and supporting all the other women participants with what they do. During the circles we discussed topics including negotiation, how to be your own hero, and what it takes to be a great leader. All these education materials are provided by the Lean In team and are very beneficial. They are all taught by experts through videos.

In August 2015, Lean In brought 50 active Lean In Circle Leaders around the world together in San Francisco for the first Regional Circle Leaders

Conference. We spent two days there. The first day of the conference was at the Facebook campus, the second, at Stanford University. It was a dream come true to meet Sheryl and all the circle leaders around the world, all with their inspiring stories.

The Saddam Regime never treated the Kurds well. We suffered a lot and have a tragic history. I believe it is time for Kurdish women to come together, speak up, sit at the table, support each other, and take the lead. My plan for this year is to launch a Lean In Chapter with a few of my friends in my hometown, Slemani. Someday I hope to say Kurdish youth have implemented so many great projects and we could have our own Mark Zuckerberg or Steve Jobs here as well, if there were investors supporting their ideas. We already have successful Kurdish women, some of them are fighters defeating ISIS, some are business women, activists, public figures, and some are illiterate, but this does not stop them: they use their talents to work day and night to support their families.

Technology can improve so many things in Kurdistan. If we can apply it to the public sector and the government in the way it's been applied in the private sector, life will get a lot easier for everyone. Also, technology can improve Kurdistan's economy, healthcare, education, and culture. Through technology, we can introduce our culture to a broader community. It used to be that, when we said we are from Kurdistan, foreigners had never heard of us. But now it is getting better, since a lot of people are traveling outside Kurdistan and international companies are investing in the country.

But we can do a lot better if we nurture the young generation. To grow our economy, we need to invest in the young generation by sending them outside Iraq to study and learn more about new innovations, or by investing in the ideas they have, so we can grow and build local companies and products. China is a good example of creating localized technology. They have Amazon and eBay, but they also use Alibaba; they have WhatsApp, but they also created WeChat. This is what I hope for Kurdistan as well, and I am sure we can do this if we invest in our youth.

Women in my generation have access to technology to develop themselves and learn skills online. However, there are still women who lack having the information on how to use new technologies to further develop themselves. I believe international non-governmental organizations (NGOs) and companies should continue to focus on investing in women by providing training and workshops locally and internationally to continue to support the development of the technology sector in Kurdistan.

References

[1] http://www.mitpressjournals.org/doi/pdf/10.1162/inov_a_00241
[2] http://fortune.com/2015/03/03/women-led-companies-perform-three-times-better-than-the-sp-500/#m03g17t20w15
[3] http://www-wds.worldbank.org/external/default/WDSContentServer/WDSP/IB/2015/05/27/090224b082eccb31/5_0/Rendered/PDF/Economic0monit0oc0liaison0committee.pdf
[4] Dawn Elliott, Pride Jamaica Project submitted by the CARANA Corporation, 2011 http://173.203.89.141/ocs/public/conferences/1/schedConfs/5/Report%20-%20Dawn%20Elliott.pdf
[5] International Telecommunications Union (ITU), "Speech by ITU Secretary General", April 6, 2010 and Indian Council for Research on International Economic Relations (ICRIER): http://www.jdic.org/files/seminars/jdic_financial_markets_symposium_-_dr_maurice_mcnaughton.pdf
[6] http://www.un.org/sustainabledevelopment/sustainable-development-goals/
[7] AFP wire story from 4 March 2015: https://www.yahoo.com/news/dubai-build-museum-future-171711442.html?ref=gs
[8] http://www.state.gov/documents/organization/183233.pdf
[9] http://www.ibtimes.com/where-are-women-davos-only-18-percent-female-participants-2016-world-economic-forum-2270816

PART II

Men and Women Empowering One Another

2

Behind Every Great Woman, There May Be a Man

"If half your team is not playing, you've got a problem. In too many countries, half the team is women and youth."—President Barack Obama

2.1 Introduction

This chapter focuses on men who have helped create more opportunities for equality either within their companies or through large-scale efforts. Much of this success was accidental, and they found out through the process of implementing more-equitable policies that women were most positively impacted. In many of the examples highlighted in this chapter, this either created better economic opportunities for a given society overall, or increased the bottom line and productivity in a given company as a whole.

2.2 Supportive Strategies for Women Employees: Lessons Learned from Zambia

By Dimitri Zakharov

Dimitri Zakharov is CEO and Co-Founder of Impact Enterprises, the first socially conscious outsourcing company in Zambia. Over the last eight years, he has combined his background in finance and business operations with his passion for social development working with Kiva, S&P Capital IQ, and the Shanti Bhavan Children's Project in sub-Saharan Africa, Azerbaijan, and India. Dimitri graduated from NYU's Stern School of Business with a degree in Finance and International Business.

"Antiquated values such as jealousy and gender stereotyping don't belong in the modern community."

In July 2015, President Obama, during his visit to Kenya, drew attention to the issue of female employment in Africa, saying, "If half your team is not playing, you've got a problem. In too many countries, half the team is women and youth." Achieving gender equality in the workspace requires more than just proactively hiring more women. It involves breaking down barriers to resources, reducing social pressures that many women in Africa face, and building their personal confidence, as we have learned in Zambia.

Creating Jobs

When we launched Impact Enterprises, supporting female entrepreneurs wasn't the first thing on our minds. We just wanted to create good jobs. Back in 2008, our company president, Dan Sutera, took a trip to visit rural Zambia with his friend David Seidenfeld, who had spent years in the country between the Peace Corps and working on his PhD. After 10 years of growing start-up companies, two of which had cracked the Inc. 500 list of the fastest-growing companies in America, Dan was looking for his next challenge. Together, they launched Impact Network in eastern Zambia to deliver an e-learning curriculum, called eSchool 360, to rural village schools.

After a few years of building schools and educating thousands of children, Dan realized a quality education isn't of much use if there are no jobs upon graduation. In Zambia, a southern-African country of about 15 million people, unemployment for 20–24 year olds is at 59%. Those who can't find work often go into the informal economy, mostly in agriculture.

Harnessing this untapped population, in an English-speaking country with a stable political and economic climate, had to be possible, and it was outsourcing that brought the solution. Over the last decade, a movement known as "impact sourcing" has emerged in the business-process-outsourcing industry with the goal of bringing sustainable jobs to untouched communities. Thanks to the diminishing costs of technology, companies were now moving into tier-2 and -3 locations in rural India, away from the established tech megacities. They were building centers in frontier markets in Southeast Asia, Latin America, and Africa. Our goal was to bring the same potential to Zambia.

I met Dan back in 2012 while working at a financial software company in New York City. Years ago, I had worked as a director for the Shanti Bhavan Children's Project, which ran a world-class K–12 boarding school for *dalit* children—the so-called untouchable caste—in rural India. I was inspired by how the students' lives had been transformed thanks to the quality of education they received, and I was eager to apply my business background to a social venture of my own.

By mid-2013, we had launched Impact Enterprises with our first client project, becoming the first socially conscious digital-outsourcing company in Zambia. By bringing digital jobs to a developing country, we weren't just pioneering a new industry but also creating equal opportunities for talented youth.

Soft-Skill Employee Training

Setting up a high-quality service company requires a huge investment in employee training. For the majority of our employees, this job is their first time working in a formal company. Beyond just learning hard skills such as computer operations and web research, they take part in ongoing workshops around teamwork, communication skills, career planning, financial management, and healthy living. Cecilia, one of our earliest employees, told us her job is teaching her "how to be responsible, how to keep time, how to have discipline. The feeling of having some responsibilities keeps me going. Now I know, wherever I go, I have to work hard."

At Impact Enterprises, we strive to keep an equal male to female ratio of employees to demonstrate that women have an equal opportunity to succeed in the digital economy. In fact, we've found that, in the first months of training, female hires tend to be more responsible, attentive, and timely than their male counterparts. Realizing the unique career opportunity they've been given—one that few women in the country could possibly have—they don't take anything for granted.

However, despite the close and supportive community we had fostered at Impact Enterprises, our young women were still struggling. Their personal charisma and determination seemed to disintegrate in a public group setting. Growing up in a patriarchal society meant their self-confidence was deeply undermined. During company-wide discussions, the women would go mute around men and were hesitant to share their opinions, even among just their fellow females. Something had to be done.

In mid-2015, we decided to launch a weekly support group exclusively for our female employees. They quickly made it their own, naming it Ladies of Victory and Encouragement (LOVE). As Dinah, one of the group leaders, explained, "We want this to be a space where we can share our ideas and learn from one another. Building our communication skills is really important."

For their first assignment, they required every member to speak up in a workshop debate without being prompted. While nervous initially, after just one month, participation sky-rocketed and they became more engaged in the company. The self-directed group has now expanded to topics ranging from entrepreneurship and career advice to self-esteem, gender equality, and maternal health.

"I know that in the future, maybe I will be managing other people," said Catherine, one of the members. "What I learn here from the managers really inspires me." Debra, one of the newer employees, said that she felt the group brought back the energy she used to have in secondary school.

For the company overall, the workshops have significantly strengthened our workforce and enhanced our services. "It has helped in numerous ways to re-establish my self-esteem and confidence as a woman," Debra told us after one session. "When a person is surrounded by positive-minded, dream-oriented, enthusiastic individuals, life is worth living because you know you can make it everywhere."

Developing the young-adult female demographic is particularly valuable, since they can immediately contribute to the workforce with their skills and perseverance. Existing social initiatives focus either on adolescent girls, in an effort to impact them at an early age, or on working adults, who need simpler career mentorship. Few resources are available to young adults who are just out of school, and who require a higher investment to impact their lives—they are more mature and sophisticated than adolescents but lack the experience of their adult counterparts.

Holistic Support System

Empowering young women is crucial for struggling countries such as Zambia. While Kenya, Uganda, Nigeria, and Ghana have emerged as tech destinations in Africa, Zambia has failed to embrace the promise of new industries over the years. Even still, riskier ventures are usually led by men. If Zambia hopes to be competitive internationally, a holistic support system must be created with the collaboration of various stakeholders to empower women at work.

In early 2015, the United States State Department of State launched the WECREATE program to establish physical entrepreneurial community centers across Africa. Lusaka, the capital of Zambia, was fortunate enough to have been chosen, and the center provides programs for training, pitch competitions, and grant funding, while also providing family support.[1] Establishing similar facilities is critical, in lieu of the currently inadequate school system.

Providing access to digital information is an easier, less resource-intensive method of creating access to education and training. Currently, Zambia is limited in Internet bandwidth, and this results in high data costs and unreliable access.

Women must also have the standard resources outside work. Health services and family planning must be accessible so women can live healthy lives and raise their families without interfering with their careers. Early education must focus on soft skills to build confidence in girls so they can boldly tackle their careers and interact with teammates. Even family members must be supportive of their sisters and daughters so they can compete with men. Antiquated values such as jealousy and gender stereotyping don't belong in the modern community.

Unfortunately, the reality is Zambia's government is saddled with debt, and a devalued currency has attenuated the abilities of local companies. The reality is that progress is going to come from the outside, either through foreign investment or from "re-patriates," who are returning to their native country. For us to see real progress, at least in the interim, the private sector will have to take on the responsibility of multiple stakeholders—both employer and educator—as we've demonstrated at Impact Enterprises.

Overcoming Psychological Barriers

The LOVE workshops at Impact Enterprises teach our female employees skills that are invaluable to entrepreneurship and allow them to overcome the psychological barriers many of them face. I'm particularly proud of one of our members, Nelicy. She's only 19 years old but one of our most determined employees. From the beginning, Nelicy put in a lot of hard work and quickly advanced through her projects. During breaks, she was gregarious and carried a bright smile, but in group discussions, she closed up. Just by looking at her, I could tell she had ideas to share, but couldn't get them out.

Later in the year, we held a workshop on financial planning to teach key concepts such as savings, interest rates, and inflation. Rising prices are a major issue in Zambia (inflation reached over 20% in December 2015 because of a slowdown in Chinese demand), and Nelicy aptly understood the implication for her pay. The following Monday, during our weekly company morning talk, I had opened the floor to questions. After a long pause, Nelicy raised her hand and boldly requested we raise wages, eloquently explaining how inflation is impacting everyone.

She was absolutely right, and we recalculated our pay structure that week. I admire her courage to stand up for her team. When we featured Nelicy in our company blog, she told us, "Actually, the time I came to Impact I was kind of a shy person. Being here I have had to open up to people."

When I think back to that Monday meeting, I remember how the long silence before Nelicy spoke up was just as loud as her voice. It demonstrates the reality of President Obama's remark about missing half our team, and I worry what else we're missing in that silence.

2.3 The Necessity of Empowering Women

By Rahilla Zafar

"I hesitated to move forward regarding fairness for women in tech, with some self-awareness of general cluelessness regarding women's issues, and decided to avoid any mansplaining."

Craig Newmark is a self-described nerd, Web pioneer, speaker, philanthropist, and advocate of technology for the public good. He is also the founder of craigslist, the almost completely free online classified advertising site that has seen more than five billion ads posted. While no longer part of the management, Craig continues to work with craigslist as a Customer Service Representative (CSR) in what he calls a "lightweight" capacity.

Today, Craig's primary focus is craigconnects, which he launched in March 2011. The mission of craigconnects in the short term is to promote and enhance the use of technology and social media to benefit philanthropy and public service. He uses the craigconnects platform to support effective organizations working for veterans and military families, peer-to-peer giving, trustworthy journalism, women in tech, protecting American's voting rights, and other initiatives.

In 2013, he was named "Nerd-in-Residence" by the Department of Veterans Affairs' Center for Innovation in recognition of his volunteer work with the department to enhance services to veterans.

Born in Morristown, New Jersey, in 1952, Craig received his bachelor's and master's degrees in computer science from Case Western Reserve University. Craig communicates regularly through his own blog on craigconnects.org and through the Huffington Post, Facebook, LinkedIn, Medium, and Twitter. He also travels the country speaking about issues, appearing on behalf of organizations he supports and delivering his craigconnects message to audiences nationwide.

Craig has been a strong advocate for the economic necessity of women in both government and as business founders. He collaborates with several organizations, including WomenWhoTech, Black Girls Code, Black Nerds Girls, and SpringBoard Enterprises.

In the following interview, he highlights his personal journey in moving forward to advocate for gender equality in the workplace and government. He also offers advice to female founders and his observations on how to leverage opportunities from the evolution of social-media platforms.

When did you start actively advocating for women in particular, and what inspired you to speak out more publicly on such issues?

It's been a very slow, gradual process, starting from a position of nerdly naiveté and its follow-through, from "treat people like you want to be treated" to the notion that everyone should be treated fairly. Seriously, from a nerd's perspective, there's no reason to treat people differently based on gender, race, and so on.

That became significant when I lived in Detroit. It was the mid-eighties. I was working for IBM with General Motors, and trying to encourage people with no tech background to LEARN to become "system engineers," a title I shared. Around that same time, that's when I started volunteering at a women's shelter called HAVEN, though I wasn't resilient enough to counsel.

In the last five or so years, people have been kind enough to help me learn of my relative cluelessness about the real-life concerns of women in governance and business, and so I figured I'd further practice what I preach, and preach what I practice (thereby acquiring as much as half a clue).

However, I learned a bit about boundaries at HAVEN. For example, I was allowed in only with an escort, for security purposes. I hesitated to move forward regarding fairness for women in tech, with some self-awareness of general cluelessness regarding women's issues, and decided to avoid any mansplaining.

Then suddenly I realized that one of my advisors, Allyson Kapin, had been running WomenWhoTech, and asked her to suggest smart ways where I could help without (much) blundering.

Since then, we've done three successful Women Startup Challenges. These have reached more than 1,000 women-led start-ups from across the country, all of them focused on solving real-world problems for people, businesses, and the world. And there will be more to come. Also, I've gotten involved with the Women in Public Service Project, helping women acquire and hold positions of public power across the world.

What advice do you have for women who are seeking investment for their companies?

My narrow perspective is from good intent, and some self-awareness of general cluelessness, having seen that in my male human peers:

- Assume that's the situation, and find the gentlest way to tell your audience that your perspective is different, perhaps the majority view, and suggest they view things in that light.
- Also assume that others are competing for attention and resources from potential investors, and a lot of people have wasted investors' time, over-explaining, telling them what they already know, and taking a long time to get to the point. To really get the attention of a venture capitalist or a banker, show respect for their time, get to the point, and don't tell them what they already know.
- Do your research on venture capitalists who are already funding your type of start-up. You can find this info in online databases such as Crunchbase by TechCrunch. Start networking with investors before you pitch them, so you already have a connection to them.

In what ways do you see social media being a tool for female founders specifically?

Social media can be used broadly to change expectations regarding a "new normal," in that gender should play no role in business decisions. Women and men should post, now and then, in a way that clearly demonstrates the new-normal attitude. This approach would've taken a generation prior to social media.

However, I think I'm seeing the new normal beginning to happen over the course of 5, maybe 10 years. That requires people of good will to say something, not in a pedantic or patronizing way, just quietly standing up for the right thing.

2.4 How to Attract (and Keep!) Mothers in Your Company

By Rahilla Zafar

Davis Smith is the founder and CEO of Cotopaxi, an innovative outdoor-gear brand with a social mission at its core. Prior to that he founded Baby.com.br, Brazil's leading e-commerce retailer of baby products. Baby.com.br was named Brazil's "Start-up of the Year" in 2012 and raised $40M+ in venture capital. Davis was recently appointed to the United Nations Foundation's

Global Entrepreneurs Council (GEC). Over the next two years, the GEC is specifically focused on developing an entrepreneurial mentoring platform and program to assist entrepreneurs in the developing world.

"It troubles me to see how badly venture-capital investing is skewed towards men. I don't know all the solutions to that, but I know one way I can help is by ensuring women hold leadership roles in my start-ups."

With baby.com.br you built a company with a large female consumer base and women as part of your executive team. How did this come about?

Firstly, the reality is that nearly one hundred percent of economic growth is coming from small and medium-sized businesses around the world. I think it's really important to include them in the conversation and to make sure they're thinking about gender issues as well, and not limit the conversation to large corporations. With baby.com.br, we felt we had an opportunity to do something different. I remember when we were first talking to one of our early investors, Kevin Efrusy, who had made the first institutional investment in Facebook, from Accel partners. He said we really needed to identify a mission. He said that, with Facebook, it was about moving fast and breaking things. It was basically, don't worry about whether it's perfect—just do it really quickly, and if it breaks, that's fine. You can go back and fix it, but learn to do things fast, and test them.

I had a couple of young kids and most start-ups were working until 10 or 11 at night and on weekends. Some my friends who had start-ups would brag about how their team worked the whole weekend or late into the night. I had no interest in creating a culture like that in my company. I wanted to go back and see my two daughters and do the bedtime routine, read them books, and have dinner with them. We were leaving the office at 6 p.m., which for Brazil was really early, especially for a start-up.

We saw that our culture attracted a certain type of employee, including a lot of women that were coming to us because they saw that it was a place that supported mothers. We doubled down on it and made sure that this was something that was really built into the company culture. I remember one of our most senior hires was someone who had worked at a major multinational for years, overseeing all of their operations and logistics. The reason she left that company and joined us was that she wanted to try starting a family and had struggled to do so for years. Six months after she joined us, she pulled me into her office and told me that she and her husband had done in vitro and found out they were expecting twins.

She cried and I cried with her, it was a very emotional, special moment. For me, it was validation that we were building a company culture that mattered. Close to 65% of our employees were female, and many were moms. I became a big believer in supporting women and in putting women in senior leadership positions. I just saw the value they added to our organization.

Could you give us a bit more detail in terms of the value that you felt they added to your organization in particular?

Our senior leadership team was half female, and something like 80 or 90% of all the people buying on our website were women. They were either buying for their own families or as gifts for friends that were having babies. Ensuring we had women at the highest levels of the company was critical. We needed and valued their perspective. I don't even have to explain why that would make more sense than just having a bunch of men sitting in a room trying to decide how or what women are going to shop for. Women lead our merchandising, purchasing, planning, customer-experience and logistics teams.

What is your background?

I moved to the Dominican Republic when I was four. I lived there for a number of years and then I moved to Puerto Rico and later to Ecuador. My family moved back to the United States when I was a teenager, but I went back to Latin America for two years as a Mormon missionary in Bolivia. Once I was married, my wife and I moved to Peru, then the Cayman Islands, and a number of years later we moved to Brazil. My childhood living abroad had shaped me, and I loved living in different countries and learning to understand different cultures and languages. I had actually never been to Brazil until I started the baby business. I looked at Brazil as a place that was really interesting with 200 million people, plus everyone was getting online in a large merchant middle class. I felt like it was a great business opportunity, especially as a dad.

I knew the founder of diapers.com from a number of years before. I had kept in contact with him and understood the model and it just clicked. I knew that the model would work in Brazil. When you're in the middle class, you either don't quite have a car yet or if you have a car in the family you probably have one car, and often that meant that a husband was going to work with the car and a new mom was staying at home. In a city of twenty million people, getting to the nearest Walmart to go buy baby products using public transportation in a city such as Sao Paulo can be a three-or-four-hour ordeal.

Giving mothers access to products they needed for their families on the Internet where orders would be delivered in 48 hours, was a solution to a real problem. We started working on the business in May of 2010, and three months later I went to Brazil for the first time. We started building the team and finally launched in 2011. In December 2013, I decided to come back to the United States to found Cotopaxi.

You speak about how Cotopaxi gave you the opportunity to include a social aspect in a business at the early stages—can you elaborate on that?

Yes. Having grown up in Latin America, I always had this huge passion for helping people and looked for ways in which I could have an impact on people's lives. I wanted to find a way to use business to do well and have an impact. With the baby business in Brazil, there's no way I could've said, "Hey, by the way, I want to start using 10% of all of our profits to give away" or "Two percent of all of our revenue is to now go to people in need."

If the social mission isn't built into the business from day one, it makes it very difficult to make that commitment as the business scales. It's much easier to do when it's built into the fabric of the business from the very beginning. The day I came back to the United States from Brazil, I incorporated the business, and I flew straight to Silicon Valley to start fundraising.

Could you elaborate on what you learned from baby.com.br and what you've carried over into your new company?

One thing that I found out in baby.com.br, with moms on our team, was that they are as efficient and productive as any other team I knew in Brazil. One thing that really surprised me is that even though we ended the work day at 6:00 p.m., while most teams in Brazil ended their workday later, is that everyone got their jobs done so efficiently. That was a huge takeaway as I started Cotopaxi. We're an outdoor-gear brand, so I really felt it was important to make sure that my team had time to go spend in the outdoors to test the gear.

Could you tell us about the UN commission you're on and the gender dynamic it has?

At the end of 2015, I was appointed to the United Nations Foundation's Global Entrepreneurs Council, or GEC—a council of eight members who come from all over the world and have a background in leading and driving organizations, typically as entrepreneurs, and who have a history of philanthropy and of giving back.

The group is very global and includes individuals from India, Puerto Rico, Sweden, the United Arab Emirates, Kenya and Uganda. It's a diverse group, and of the eight members, five are women and three are men. This is the first time the council has had such a high percentage of women and it's also the most international the group has ever been.

We have two focal points for the next two years. The first is to help create ecosystems for entrepreneurs and a platform where entrepreneurs in the developing world can mentor and be mentored. Within that, we're primarily focusing our efforts on youth and women. We know that these groups have the highest impact since women have been shown to spend their incomes on their children, whereas men are more likely to spend it on themselves.

The second area that we're focusing on is on refugees. The chairman of this council is actually a former refugee himself from Rwanda. He has a really unique story, and it's an area that a lot of us on the council are passionate about. We're focusing on building more awareness around the growing global refugee crisis and assisting refugees at every stage—from immediate aid as they flee their home countries to resettlement assistance.

Having successfully raised funds yourself, what are your feelings on women having access to investment?

I'm the father of two girls, and I would love nothing more than to see them become entrepreneurs. It troubles me to see how badly venture-capital investing is skewed towards men. I don't know all the solutions to that, but I know one way I can help is by ensuring women hold leadership roles in my start-ups. We know that a lot of start-up founders come from start-ups themselves. They go build experience working at a start-up, and then leave and do it themselves with the right networks in place. As a start-up founder, I'm responsible for making sure my team is building the next generation of entrepreneurs and that we include minorities and women in that growth.

This takes real effort because the reality is that without even realizing it, humans have biases. We tend to hire people that have experiences like us that maybe look like us, without even thinking about it.

From the early stages of a business, we have to be thinking about these things. You can't just think, "Oh, as we grow or as we scale we'll be able to get more diverse or we'll think about those issues later." I think it's something that you have to do from the very earliest stages of a start-up.

Our board at Cotopaxi is half male, half female. Our biggest investments are led by female partners in venture-capital funds. It's something that I'm pretty passionate about, and that we think about constantly.

2.5 NA3M We Can: A Saudi Prince Roars for Women Gamers . . .

By Rahilla Zafar

"When parents see and support their daughters' creative passions and skills, they'll find that this is something that benefits the whole country, not just the family."

Video games are a multi-billion-dollar industry, with the global gaming market projected to reach over $113 billion by 2018, according to the analyst firm Newzoo.[2] In the Middle East and North African regions, the video game market is burgeoning, with an expected annual growth rate of 29%.[3]

NA3M, based between Jordan and Germany, is one company receiving a lot of attention lately. Its CEO and Founder is from Saudi Arabia, and is an award-winning pioneer in the tech industry. His name is Prince Fahad Al Saud.

After graduating from Stanford University and prior to launching NA3M, he was hired by Facebook in 2008 to help launch the popular social media site's Arabic version. NA3M is interesting for a lot of reasons, and not least because more than a third of the staff are women. Prince Fahad has made it one of his primary missions to advocate for equality through technology by launching a new media platform that champions, fosters, and celebrates Arab identity—with a focus on empowering women and girls.

When asked about his motivation behind launching NA3M, particularly when it comes to helping women to shatter the glass ceiling in Saudi Arabia, Prince Fahad told Talk Media News, "This is my responsibility. As someone who was raised by strong women, I want to represent them through the work that I do. SGR (Saudi Girls Revolution) isn't just our latest piece of animated content, it's an animated example of female power, super power actually."

In fact, Prince Fahad attributes much of NA3M's games to his team's ability to empower women through tech-driven tools that allow them to reach and connect with the outside world, shining a rather overdue spotlight on their efforts.

Hiring women is not just about empowerment and igniting cultural change, however. Prince Fahad and the gaming industry in general also understand that women make up the largest group of consumers (approximately 60 percent) of mobile games and apps. Having a woman's perspective is important for business growth.

Female Employees on Work, Gender, and Cultural Change

Abeer Ahmad, originally from Palestine, studied computer science in Amman but did not receive any specialized training in gaming. In fact, people told her that she wouldn't find work in the programming field and that gaming wouldn't be successful in Jordan. They were wrong on both counts.

Despite her lack of training in game design, Ahmad made a game for her graduation project and received high marks for it. Subsequently she found an opportunity to interview at NA3M, where she has worked since 2013. She loves her work and says that game design provides especially exciting challenges because each game is different from the one that comes before it, and staff are constantly learning on the job.

Her colleague Sarah Kilani says, "When I applied here, and they saw my work, they told me that I didn't think highly enough of myself: 'You're much better at this than you think you are.' That was a shock for me! I'm not used to hearing things like that, except from my parents."

She credits her family with the happiness and success she has found at work and feels that the support of family is essential for a woman's achievement. "When parents see their daughter's passion and skills and support them, they will find that this is something that can benefit the country, not just the family," she says.

Kilani is from Jordan, where she studied in the University of Jordan's visual arts program. When she was in high school, however, she focused on science. While people were originally confused by her transition to art, both skill sets have served her well at NA3M, where she creates game concepts and visuals.

Another employee, Jude Soub, feels there has been tremendous progress in gender equality not only culturally and in the home, but in the workforce. She sees greater equality in the Jordanian workforce than in the United States, where she studied and interned in public relations for several years and where she recalls hearing many colleagues complain about unequal pay between men and women—something she says is not an issue in Jordan.

Soub is also confident that education for both men and women in Jordan is a priority and well advocated for by Queen Rania. "I don't think education has ever been an issue here," she says. "I think everyone here strives to be who they want to be. Education is always number one in Jordan."

Rana Romoh is newer to NA3M, where she provides copywriting and digital marketing for the company, including translation, social media, search-engine optimization, and online advertising. She studied English Language and Literature at the University of Jordan, and wanted to get into marketing ever since she graduated. She attributes her professional interests in part to her brother, who is a successful digital marketer for a large company. "He is successful. Why not me?" She asked herself.

Before working for NA3M, Romoh had settled for a public-relations job in Dubai. When she moved back to Jordan, she learned about the gaming company and applied immediately. At NA3M she has found a lot of freedom to innovate and to grow her skills through advanced certifications.

All four women see a dramatic and positive change in society among women, especially when they look back across the generations. Women are choosing careers, waiting to have families, and pursuing their ambitions. Although there remain cultural norms worth challenging, these women feel strongly that, while the main interest of most companies is making money, Prince Fahad and his team at NA3M are dedicated to achieving more than that. Their goal is to continue to lead a significant change in popular culture, technology, and the workplace, both locally and globally.

2.6 Female Leadership and the Future of Water Security in the Middle East

By Rahilla Zafar

"If there were more women involved in high-level decision-making, the water sector, sanitation, and water supply would receive much greater policy attention."

For nearly a decade, Kate Rothschild has been actively involved in her family's philanthropic trusts, including Yad Hanadiv, which operates in Israel. Rothschild helps oversee grant approval across the sectors of the foundation, which includes initiatives around supporting Arab society, academic excellence, education, and the construction of a new 21st-century state-of-the-art building for The National Library of Israel in Jerusalem.

Under her leadership, Yad Hanadiv established its first ever department focused specifically on environmental issues including a mapping exercise identifying marine areas that impact tourism, fisheries, and recreation. With water scarcity in the Middle East North Africa region being one of the most severe environmental issues in the world, EcoPeace Middle East (formerly Friends of the Earth Middle East) was an organization that caught her attention because of their tri-lateral approach being run by Jordanians, Palestinians, and Israelis. When environmental issues such as water are being addressed, Rothschild explains, "borders become irrelevant and cooperation becomes essential."

It isn't just a collaborative approach across borders that EcoPeace focuses on, but also gender. Recently it established a new program specifically centered on cultivating young women to enter the water engineering field and engage in civic leadership. Rothschild joined EcoPeace's advisory committee and traveled to Gaza with the organization's Israel director Gidon Bromberg. She says, "I was profoundly moved by what I saw in Gaza, on the one hand by the bravery and strength of the very normal people living there, but also by the depth of the humanitarian crisis. I also realized that people have no idea what's actually going on in Gaza in terms of the environmental crisis, which will continue to impact the entire region if not addressed."

She works closely with Bromberg, to help draw international attention to the water and sanitation problem in the region, with a focus on Gaza in particular, while her foundation also provides financial support for their initiatives. EcoPeace is the only regional organization that brings together Jordanian, Palestinian, and Israeli environmentalists to promote sustainable development and advance peace efforts in the Middle East. It has offices in Amman, Bethlehem, and Tel Aviv, and employs 100 paid staff and actively involves hundreds of additional volunteers.

Having spent much of his childhood in Australia, Bromberg's return to Israel was driven by a desire to contribute to peace-building efforts in the Middle East. Bromberg is an attorney by profession, specializing in international environmental law, and is an alumnus of Yale University's World Fellows program.

In this edited transcript, Bromberg speaks about the environmental issues that EcoPeace is currently working on, as well as about the importance of female engagement both at the engineering and leadership levels.

What is the gender aspect of the region's water and sanitation crises and how are you trying to address it?

2.6 Female Leadership and the Future of Water Security in the Middle East

Woman and girls traditionally pay the higher costs for water and sanitation failures. If there were more women involved in high-level decision making in the water sector, sanitation and the water supply would receive much greater policy attention. We've witnessed the gender impact on women through our Good Water Neighbors program. It's a community-based constituency-building program where we work in schools with youth, with activists, and with mayors. This year we'll launch a gender-based program in Jordan, Palestine, and Israel, which will focus on how to empower girls and women to be more educated and more involved in the water sector. We're going to showcase existing female water engineers and have them speak in classrooms so that they can be role models and encourage other girls to be water engineers. And then we're going to give scholarships to the most dedicated young female water trustees.

We're developing programming for the whole school, and not just for girls, related to the importance of involving women in policy and how current policies don't fully take into account the issues and interests women see as priorities.

Our youth programs work with 60 schools in total; 50 students from each community go through the curriculum with us every year and those students then present their projects to the whole school. It includes school fairs, environmental walks, and engaging the parents, which is essential because through parents you engage policy-makers and mayors in particular.

When you don't have proper sanitation at home, the burden often falls on girls and women. When there isn't proper sanitation in schools, girls often have to stay home. While boys can go out to a field to urinate, girls can't. And girls who are having their periods will also need proper sanitation facilities. The lack of such facilities, then, will mean that a lot of girls will simply not be sent to school. The implications of not having proper sanitation or a sustainable water system in place are just so far-reaching. Therefore it's really critical we invest in water education for both girls and boys so that everyone can understand the importance and relevance of female leadership in this sector.

What are some of the areas in which EcoPeace Middle East is currently engaged?

The Good Water Neighbors project, which has been running since 2001, is based on sets of cross-border partnering communities sharing a common water source, thus promoting environmental awareness and peace building.

This is a water and health security risk not just for Gaza but for Israel, too. The concern is that pandemic disease could break out in Gaza because of the poor water quality and sanitation issues, and together with the sewage, that type of disease could have a horrendous impact in Gaza, Israel, and Egypt.

Water security is now also at risk. Desalination plants are designed to separate the salts of the sea and other minerals from the sea water, so if the water is contaminated to begin with then it remains contaminated after the desalination process. So that's why the Israeli desalination plant was closed, because the intake was polluted seawater.

How could technology help enhance your work?

In order to avoid pandemic disease outbreaks, the public on all sides need to better understand the costs of failing to respond to the current situation in Gaza. There's a lot of public awareness that could be promoted visually through technology. We want to get a hold of satellite imagery and maybe infrared to see how the sewage leaving Gaza is moving, impacting the beaches in Gaza itself and on the Israeli coast. We want to distribute kits to the general public where they can undertake tests themselves and share those results through an app—to self-analyze the seawater, where they find pollution, what they find it and at what time, and thus create a crowd-sourced pollution map. We want to raise awareness that this is a ticking time bomb. We'd like to get out on a boat and on a helicopter as well, to film the sewage leaving Gaza so it could be far more visible, and share the technology on Facebook and Twitter, so real-time images could become available.

What are some of the ways in which you've been able to raise awareness so far?

What we're seeing in the public on both sides is that people remain ill-informed. Had we not put in a request through the Freedom of Information Law in Israel (which gives the public the right of access to all types of records held by public authorities, with some exceptions), the public would never have known that the desalination plant in Israel was closed because of the sewage in Gaza.

So clearly, the government is not willingly sharing this information; the Israeli water authority has not announced it to media or to the public. We have a responsibility to continue to empower the public and the individuals concerned, and technology will play a key role in doing that. We need to use technology to understand and visualize pollution levels, and then share the information as widely as possible.

There's been tremendous over-pumping (over-drawing) of the Gaza aquifer by the Gaza population. Imagine LA surviving only from the water underneath the city of LA—the city wouldn't survive for more than a few months. That's been the situation for Gaza now for several decades. Gazans are pumping water from underneath their feet and they're drawing three to four times the renewable rate, which means the level of water in the aquifer has been dropping significantly.

2.7 Shifting the Gender Paradigm in Saudi Arabia: Thousands of Women Enter the Workforce

By Rahilla Zafar

Glowork is a female empowerment organization that has created thousands of jobs for Saudi women. Founded in 2011 by Khalid Alkhudair, Glowork has been recognized globally for providing the best innovation for job creation by the UN, the International Labour Organization (ILO), and the World Bank. Alkhudair is an active writer who contributes to many international publications and has received numerous entrepreneurship awards.

"We found that everyone in Saudi Arabia wanted to hire women, but they didn't know how or where to recruit them."

When Khalid Alkhudair founded Glowork, he set out to solve a problem: to not only help Saudi women find jobs, but to empower them to succeed. He and his team worked at the grassroots level to create a clear strategy—an approach he credits for their success. They researched what women wanted, why the concept of women in the workforce was so challenging for Saudi society, and what was really happening on the ground.

What they learned surprised them.

"We found that everyone in Saudi Arabia wanted to hire women, but they didn't know how or where to recruit them," he said.

That's where Glowork came in. It began in 2011 as an online portal to link employers with female job seekers. The organization has created thousands of jobs for women in Saudi Arabia. Its database includes 2.2 million women who have signed up to access the many services available from the portal.

Users can search and apply for jobs, receive notifications of new jobs, and get updates about jobs that fit their interests.

Since 2011, Glowork has helped an average of 25 Saudi women a day find jobs, according to Jowharah Al-Theyeb, senior manager at Glowork and the first woman in her family to work in the private sector.

"We work with women from all different levels and skill sets," she explained. "We offer entry-level positions in factories, retail and childcare, but we also offer positions for more-qualified candidates from graduate programs. These people have been hired at companies such as Microsoft, Oracle, and Cisco."

Each year, Glowork holds the largest female-only job fair in the world in Riyadh. It attracts about 55,000 women—and for good reason. Glowork works with every university in the Kingdom, about 270 companies, and numerous NGOs, all of which are represented at the fair.

Cultural Obstacles and Legal Barriers

When Glowork started its work in Saudi Arabia, the team faced an uphill battle.

"It was a very slow transition toward acceptance of more women in the labor force. It was so new that people didn't know how to react to it," Alkhudair said. "People questioned whether it was right or wrong, whether it followed religion and tradition. But at the end of the day, I think people saw it was positive for the economy and positive for themselves and their families.

"Glowork has given Saudi women a space to develop themselves, to grow and utilize their talents."

Their challenge went beyond just recruiting and training female candidates, or even working against cultural and generational beliefs about women in the workplace. There were a number of problematic laws and workplace obstacles in place.

Transportation is one example. Women are not allowed to drive in Saudi Arabia. Al-Theyeb says this is a major reason why women don't seek employment—or have trouble showing up. Some employers in industrial cities and factories provide transportation for female employees. Others may offer transportation allowances, but they don't often cover the full cost of getting to and from work.

When Glowork first entered the Saudi job sphere, Alkhudair and his team found a lot of gender segregation: separate buildings, separate doors, separate everything. Today, it isn't as extreme as it used to be. A man may see women in the office, but women have their own space to work, eat, and pray.

"The old model carried a lot of costs to employers," he explained. "If I were to hire two or three women, I would need a whole building by itself. It wouldn't make sense to hire just one.

"We worked with these limitations and tried to be innovative. We created a virtual office system to allow women to work from home. And in rural areas, it enabled both men and women to change their perceptions about what it meant to be employed and working."

There have been a lot of positive changes, too. Alkhudair credits the government for being very helpful in overcoming some of the biggest challenges. It helped create subsidies and launch awareness campaigns. Maternity leave—and paternity leave—have been extended. If a company employs more than 50 women, it is required to have a nursery within the organization.

But the most impactful change it made was overturning a key law: a legal ban on women working in retail.

Previously, women would buy most of their products online, especially personal items such as lingerie and cosmetics.

"There was this sense of shame in buying these products from a man," Alkhudair said. "When the government removed that barrier in 2011, at some retail outlets, sales increased by 300 to 400 percent."

Navigating the New Workplace

As a result of the shifting tide in the Saudi workplace, many men and women are having their first interactions with the opposite sex outside of their family. For many of them, they have had limited interaction with the opposite sex their entire lives. There's no place to publicly mingle, sit down and get a coffee. In their society, these types of basic social interaction do not happen.

Alkhudair describes the changes unfolding there as "living in a social experiment."

"You can imagine how challenging this change is for both sides. They don't understand each other. Expectations are different and they don't know how to have a conversation," he said.

"It will be very interesting to see how these changes develop over the next several years. There has been a lot of progress in a short period of time. As companies adapt, women are adapting to the work environment."

But it isn't always very smooth. Miscommunication is a big issue for both female employees and their employers as they learn how to navigate the workplace.

Al-Theyeb offers an example of a woman working at a retail store in a mall who was doing very well, but suddenly resigned without warning.

"The employer called and asked us for her feedback. When we called her, she told us that the restroom facilities for females were not clean. She didn't know how to communicate it to them. She preferred to resign than to tell her manager that the restroom was not clean."

Glowork holds workshops for employers to educate them on how to treat their employees. They also hold workshops for employees to teach interviewing and communication skills.

Employers and employees aren't the only ones adjusting to new roles. In many cases, women need permission from their family to interview and work. Al-Theyeb estimates as many as 60 to 70 percent of women who want to join the workforce may be prohibited from doing so.

She has seen women come to interviews who need to step away to call their husbands to double check answers. Or, their husbands may be waiting outside the door. Resumes come in with mobile contact numbers on them, but when Glowork tries to follow up, they reach a husband, father or brother.

"We had this one lady. Her father would watch her outside the store she worked on a daily basis. He wanted to check on her and see what she was doing. I think he did it for the first week," Al-Theyeb said. "Now, I believe she's a store supervisor."

Women in Technology

Alkhudair sees a lot of potential for women to enter the technology sector. There are a lot of female graduates in IT, but not enough companies have changed their mindset about hiring women as, say, network engineers.

"That will require time and education, but there's not much potential there."

Some of Glowork's partners are already providing these opportunities. Al-Theyeb cites companies, such as Price Waterhouse Coopers and Accenture, who hire a lot of female network engineers.

The tech industry is a space where two types of women have an advantage: younger tech-savvy graduates and those who studied abroad—an especially coveted group.

Many employers demand technology skills. Glowork works with many start-ups working on an app or a website. Many grad students already do this type of work as freelancers, so they are an easy match.

Still other companies prefer to recruit from the pool of thousands of Saudi women fully funded to study abroad by the government. Through the Saudi

Embassy in the United States, Glowork is working to build relationships with universities to help attract those candidates.

Virtual Office Tools

Glowork turned its online portal into a huge opportunity for thousands of Saudi women who live in remote areas: It created a virtual office, which allows them to work remotely.

"Most of the people who are working from home are from smaller cities or rural areas so they don't have a lot of opportunities," Al-Theyeb said. "They are educated. They have good skills and they are very qualified for the positions. They are very eager to learn and they want to prove themselves."

Here again, communication skills can be weak and present challenges.

"Because they are working at their homes, they only communicate with their husbands, their families, their children and so on," Al-Theyeb said. That presents a unique challenge as they not only struggle with communicating effectively, they also lack the ability to hold a face-to-face conversation.

Glowork offered training on how to communicate, on customer-service skills, and on what to do when common problems arise.

Matching Training with Recruitment

Al-Theyeb says one of Glowork's biggest mistakes early on was focusing so heavily on recruitment.

"We thought that was the only problem," she said. "Now through experience, we know and expect that people need training. We are trying our best now to start training projects. I think if we can match training with recruitment, we'll get a better result."

Glowork offers vocational training and career counseling, as well as a pilot program tailored to training young adults entering the job market.

With funding from JP Morgan, Glowork has partnered with the International Youth Foundation to offer the Passport to Success program for young adults aged 18 to 25. It offers training in industries from hospitality and administration to health care and retail.

Participants rotate between training and on-site work experience. They may attend training sessions four days a week, and work two days. Glowork measures their sustainability and whether they stay on the job for three to six months.

Opportunities are emerging in other areas. GE and Saudi Aramco have jointly launched the first all-female business in the country. Run by Tata

Consultancy Services and based in Riyadh, they are hiring 3,000 women and will train them on IT, support functions, HR and administration. These are roles that were previously outsourced. Glowork has already worked with them to hire about 300 over the last several months.

What's Next?

Alkhudair plans to expand the Glowork concept globally. The company has already launched a new app that can be used anywhere in the world.

Within Saudi Arabia, the company can point to many success stories. Besides countless individual successes made possible through its services, Glowork has managed to effect change so women can find opportunities without looking for work abroad.

"I know for a fact that we're moving very fast, and positively," Al-Theyeb said, referring to Saudi society. "In five years, you will see a big difference. A lot of youth now have great opportunities ahead of them."

2.8 Governments, Corporations, and Paying Women to Learn

By Brian Rashid

Brian Rashid is a professional speaker, author, and trainer. He travels around the world speaking and writing about how to become a master storyteller and public speaker. He also teaches people how to achieve financial freedom by starting businesses or creating additional sources of income doing things they love.

In the United States alone, 46 million Americans are living in poverty.

For the past 15 years, I have been working with people that fall into this category. Ten years ago, during my second year of law school at the City University of New York, I participated in a clinical program called the Economic Justice Project (EJP). Our goal was twofold. First, we represented single mothers within the CUNY system who had had their public assistance terminated without cause. They relied on this money to continue their education while putting food on the table for their families. Second, we lobbied legislators in Albany to take a more inclusive stance on what constituted the "work hours" required of recipients of public assistance in order for them to continue receiving aid. For example, instead of requiring women to rake leaves in Central Park, or pick up trash in the subway, we pushed for homework time and externships to count toward the requisite hours.

Ten years later, I took a step back to realize what happened during that semester. The government was paying women to learn, and the results were positive for both the students and the government. For many of these women, graduating from college was on ticket out of poverty. According to a University of Pennsylvania Law Review article, "Poverty Law and Community Activism: Notes from a Law School Clinic" by Marilyn Gittell, 90% of women who graduated from college never went back on welfare again. The money kept them in school and was important. One woman, whom I'll call Diana, was a single mother living in the Bronx. She had recently left an abusive relationship and relocated with her daughter. Diana received welfare, went to school full-time, and obtained a bachelor's degree. A decade later, she has not needed a dollar of welfare money, and has been a full-time nurse in one of New York's best hospitals. At a hearing during her last year of college, I represented Diana when the government tried to cut her benefits without notice. Under the Due Process Clause of the 14th Amendment of the United States Constitution, a notice and a hearing are required before benefits can be discontinued. We won the case, and to this day she credits the completion of her schooling to the fact that she had the financial support that the welfare provided her with.

This caused me to ask myself why governments or other entities around the world are not funding the education of women. Welfare is just one opportunity, and it is far from perfect. The system requires you to jump through constant hoops, stand in long lines, and do unrelated work. Another opportunity is to be paid for learning-skill sets that will help you with gainful employment or the tools needed to start your own company.

As Tess Posner, the managing director of Samaschool, points out in Disrupting Unemployment, work offered through online platforms is growing by 22% each year, compared with a 3% growth of offline jobs.[4] By 2025, online work will add $2.7 trillion to global GDP and enable 54 million people to access jobs. This is great news for people such as Diana, but there are two main problems. First, as many as 200 million people around the world lack access to the basic skills required to participate in the global digital economy and earn a satisfactory wage. Next, many people are unaware of the opportunity the online economy presents.

Luckily, there are a number of programs that provide the training needed to compete in this new and global economy. For example, Samaschool is a non-governmental organization that offers a 10-week, 90-hour bootcamp that teaches students the fundamentals of how to succeed in online work and prepares them for in-demand jobs through specialized tracks and certification.[4] Samaschool offers this service at no charge to its students in order to make it accessible to even the most vulnerable populations. The advantages to the women who use this platform are only part of the benefit. Non-profits, corporations, and the government can also enjoy the benefits of paying vulnerable populations to learn.

The connection between non-profits and corporations can be mutually beneficial. A corporation wants to acquire customers and make money. Non-profits can accelerate those two goals. Samaschool is a great example offering its courses to students at no cost. Relying on grants to keep its doors open, this model does not allow them to profit. Imagine that a non-profit such as Samaschool sold its courses to banks, and the banks offered it to their potential customers as a service. What if Chase Bank said, "If you sign up with us today, we will give you a free 10-week training at Samaschool so you can develop the skills needed to get meaningful work." The bank's incentive here is to increase their customers' earning power. The more money bank customers make, the more they invest. When they have made enough to buy a home, they will need a loan from their trusted bank. The same bank that gave them the tools needed to earn the money that allowed them to fulfill their American Dream in the first place. David Nordfors, whom I cited above, recently suggested this to the banking world, encouraging them to invest in innovation for jobs. Once the banks start competing to find the best "nonprofits" to partner with, then we are on the right track. And once companies (in this case, banks were our example) start contracting non-profits for these types of services, the shift into a for-profit mode will follow.

Next, the government should use some of the $35 billion spent each year on unemployment and pay people to learn. Diana, from our example above, was able to pull herself out of poverty because of the skills she learned in college while welfare covered her basic expenses. The government's investment in Diana paid off, as she never returned to the welfare rolls again. For those in the world who do not have access to college or who have to work in jobs that are not building skill sets that provide financial freedom, services such as Samaschool become critical. Imagine the single mom who is paid a monthly stipend to get training on Samaschool's platform as a virtual assistant, a graphic designer, a copywriter, or any other number of the in-demand freelance roles. This takes them off the welfare system, and provides confidence that real job stability is within reach. You are no longer a number in bureaucratic hell. Rather, you are an appreciated member of a team with skill sets that allow you to start earning independent of government funding.

The tools that are needed for this progress are in place. The democratization of computers and the Internet makes this concept of mutual benefit attainable. Diana, or any other woman living in an abusive relationship, could not learn if she had to pay thousands of dollars she does not have for a computer or rely on a slow modem to connect to the Internet. Neither constraint applies any longer: today, you can get a computer for a few hundred dollars, and Internet connections are only becoming faster and more reliable across the globe. What's more, the democratization of the computer and its connection to the Internet marks the end of the "unemployable outcast." Twenty years ago, there were hundreds of professions that would never have appeared when you did a google search. Had anyone heard of a professional ethical hacker 20 years ago? No. How about an app developer? Nope. This is funny to even write, considering I meet a new app developer every time I go out in San Francisco. The computer and the Internet not only allow people to learn the skills needed for these professions (online, for free or at a low cost), but also provide access to others looking to hire these specific services and skill sets from every corner of the world. Age does not matter. Physical location does not matter. If you have a computer and access to the Internet, you are in business.

Paying women to learn can look a number of ways and benefit a number of different organizations. But one thing is for sure. When you pay women to learn, everyone wins.

References

[1] http://www.state.gov/e/eb/cba/entrepreneurship/gep/wecreate/
[2] https://connect.limelight.com/blogs/limelight/2015/08/31/consumers-as-producers-how-games-and-video-converge-to-drive-growth
[3] AlJazeera, Nadine Ajaka, 15 June 2014: http://www.aljazeera.com/news/middleeast/2014/06/jordan-videogame-industry-reaches-ps4-201467131741514307.html

PART III

Bold Leadership: Women Changing the Culture of Investment and Entrepreneurship

Introduction

This section provides insights from successful female investors and founders. On a policy level, Terry Reintke, the youngest female member of the European Parliament, initiated a 2015 report on gender equality and empowering women in the digital age on behalf of the Committee on Women's Rights, for which she is the rapporteur. When speaking to Reintke, she points out that most of the funding the EU provides for digital development goes to men. In Bulgaria, she notes that the digital sector makes up 10% of the country's GDP. She continues to call for the EU's digital agenda to take gender inequality into account, and advocates for more programs offering seed capital to women. "These funds are used as a policy tool to support certain sectors, and we want to have more women in technology and ICT overall. In the digital arena, we need to make education more accessible, and we're still battling with gender stereotyping," she explains.

Many advertisements for such funding and training programs target men, leading Reintke to say that more work needs to be done to reach women so that they, too, can have start-up kits that can help them start a company. "It's not just about money—it's about training, too. If you don't have a properly functioning network or access to certain knowledge or experience, you're at a disadvantage," she says. In the follow chapters, we hear from female founders and investors on how they found success and what additional work needs to be done.

3

Targeting Untapped Markets

3.1 Creating an Ecosystem for Black Founders

By Rahilla Zafar

"Even though African-American women are the fastest growing group of entrepreneurs, they raise on average $36,000, which is not really enough money to do much of anything with."

Monique Woodard, a venture partner at 500 Start-ups, established Black Founders in San Francisco in 2011 with her friends Chris Bennett, Hadiyah Mujhid, and Nnena Ukuku. At the time, there was no community in Silicon Valley we knew of for black women working in technology. Each often found herself being the only black person in the room at hackathons and other local networking events. Then the idea came to them: what if they were able to extend the support they were showing each other to more black people working in the technology industry? Monique immediately registered the domain and put up an Eventbrite invitation to host an initial networking event. They were pleasantly surprised by the response to their idea and by the high attendance. Their concept was simple: get people together over drinks once a month.

This simple idea of creating a community grew into something much bigger. People outside San Francisco started asking them when they'd have events in places such as Atlanta, New York, and other cities. At the time, Woodard was working on a start-up with her co-founder Chris. Woodard's other co-founder, Nnena, was a lawyer; even though the three of them were extremely busy, they all decided that this was something that they had to do. They started with a conference series called Ideas Are Worthless. They held events in Atlanta and New York. Overall, people were very receptive to their ideas because they focused on helping people move from having an idea for a start-up to actually executing on that idea.

Then they started to think about the pipeline of how people get into technology. They knew that starting at historical black colleges and universities (HBCU) was the next thing to do. In their own careers they had gravitated towards San Francisco because they were really into technology. But there wasn't always a typical roadmap for people of color, or for black people specifically, coming out of historically black colleges or universities and going into a start-up or launching their own. The group started doing programs at HBCUs. They held the first hackathons at Howard University and at the Atlanta University Center Consortium.

Over the years the team has seen the community in San Francisco and in Silicon Valley expand. When Woodard first moved to the area in 2008 from Florida, she recalls hardly ever seeing another black person, and if she did, it was someone she already knew. Now that has changed with the fabric of the city becoming more diverse.

She credits the women in her network for much of her personal and professional success since her move. She references her friend Judy McDonald Johnson, whom she met at a TED conference and who has connected her to an outstanding network of black women professionals. Woodard believes in "paying it forward" and continues to connect other women to each other, especially women of color. Woodward also personally advises as many women of color in the tech industry as she can.

According to Woodard, it is particularly hard for African-American women to raise money. "The #ProjectDiane report from Kathryn Finney (and sponsored by GoDaddy) came out recently, and I think it cited 11 black women who have raised over a million dollars. Even though African-American women are the fastest growing group of entrepreneurs, they raise on average $36,000, which is not really enough money to do much of anything with."

She also points to the dearth of black women in venture capital and investment. "There's Lisa Lambert at Intel Capital, Shauntel Poulson at Reach Capital, there is me at 500 Start-ups ... literally you could sit here and name them all and it will not take you that long. So not having women in positions of power to advocate for you, or having access to that specific kind of network, definitely cuts you off when it's time to raise funds."

Woodard joined 500 Start-ups because she was thinking very deeply about what Black Founders was going to do to open up more capital for entrepreneurs. She was trying to decide whether they should raise their own funds, start a syndicate, or take a different approach. Her first meeting at 500 Start-ups was very casual, and she shared her thoughts on access to capital for underrepresented entrepreneurs.

"It wouldn't make sense for me to go to a firm that was really focusing on series-A funding, because most black and Latino founders haven't gotten that far yet," says Woodard. "The place where those founders need the most help is in the early stage and finding and raising seed capital. This made me a good fit for 500 Start-ups, which also has an approach that involves making a lot of bets. I didn't want to do five deals in a year, I wanted to write a lot of checks, and 500 was the place to do that."

The diverse culture of the company was a bonus, and the support of Founding Partners Christine Tsia and Dave McClure made 500 Start-ups an easy choice for Woodard.

Today, part of her outreach involves going to places outside Silicon Valley. Plenty of investors sit in Silicon Valley and comment that they "don't know any black or Latino founders." In Woodard's opinion, you certainly won't if you don't get outside the Valley. Being able to go to places such as New York, Baltimore, Miami, and DC—places where they have sizable black and Latino populations—is a really important way to build that pipeline.

In Woodard's opinion, not focusing on or thinking about the black and Latino communities is a mistake for any investor. The combined purchasing power of these consumers is currently $2.5 trillion in the United States alone. "By 2040, people of color will be the majority in the United States," says Woodard. "Markets are shifting and changing, and being able to find founders who understand those markets and really understand how to speak to African-American consumers is going to be really important to the continued health of a fund."

500 Start-ups has already had success investing in black founders. "Take Mayvenn, a hair-extension product, and its co-founder, Disshan Imire. No other type of founder could have started that company and done as well as he's done," says Woodward. "So I think being able to understand those founders and understand the markets that they're working in is key to continuing to make money in this space. I think there is a huge amount of money that other investors are leaving on the table and I'm happy to take it."

3.2 Solving the Technology Skills Gap through Investing in Minority Communities

By Rahilla Zafar

Laura Gomez has made repeated headlines with her suggestions to the tech industry to start thinking about diversity and inclusion differently, including:

"Hey Silicon Valley, give your Nanny's Children an Internship." CEO and Founder of Atipica, Gomez has called the tech industry out on its "tweet activism" around diversity. "Care about diversity in tech?" she asks. "Ask your building security guard, office janitor, gardener, nanny and/or housecleaner if their kids need a job, an internship, or a scholarship. Give an internship to the children, nephews, nieces, grandchildren of these hardworking folks."

Gomez can speak to all sides of this issue. Her mom has been a housecleaner and nanny for several tech executives, including some well-known CEOs. And although she was treated very well, Gomez says no one ever asked her mother if her daughter might want an internship, even though she was accepted to Harvard and Stanford and attended Berkeley, and despite her impressive roles at Twitter, YouTube, and Jawbone early in her career. This lack of awareness continues, argues Gomez, who is equally surprised by the "whoa-I-never-thought-about-it" reactions to her suggestions.

"How can you *not* think about it?" she asks. "The bus driver for Apple that lives in his van because he can't afford to actually reside here, the single mom that wakes up at 4:00 a.m. and gets home late at night to drive tech workers to work and back, the security guard at the office building ... there is this 'banalization' where you just don't see these people as human beings anymore. That to me is really scary. How can someone take care of your children, be your driver, greet you at your fancy offices, and you do not think about their potential, who they are, their families and where they come from?"

Gomez asks, "How can we talk about diversity when we can't even acknowledge or assist the diverse workforce the tech industry already employs?" Equally shocking to Gomez is that despite the positive reaction she gets from her talks and articles, there remains no obvious path for contract workers to become part of headquarters, or a scholarship fund for the children of these workers.

"My company, Atipica, is made by immigrants," says Gomez. "My team is global, and I want to create jobs here in California—it's really all about immigration flow. My story is unique because I came here as a child, undocumented, and I write a lot about this topic on LinkedIn. But—I stopped reading the comments." Gomez says while a lot of people attempt to praise her for her views, they are actually xenophobic towards immigrants, particularly Mexicans, saying "Well, you were the smart one. You are an exception." She is buoyed, however, by the positive responses from younger generations.

"When I spoke at the Udacity keynote, I got e-mails from all over the world: from Bangalore, from Nigeria, from Canada, a Filipino immigrant in

New Jersey—all of them talking about how they wanted to get into tech and they don't have the traditional background but they are inspired by my story."

Gomez worked throughout high school, 20 to 25 hours a week at a shopping center near Stanford University. She took and excelled at STEM courses in high school but feels that no one really pushed her to pursue a technical path. It wouldn't be until later in her career that she would fully appreciate her technical talents.

She didn't tell anyone that she got into several Ivy League schools because of her undocumented status and the fear of being across the country—being close to her family was an important deciding factor. "I didn't want to be influenced by the label, and I think either way I knew that my education was going to be a good experience regardless of where I went." Gomez was accepted at Stanford but didn't see people like her there, and instead, fell in love with UC Berkeley. "[The community at Stanford] was more likely made up of the people that my mom cleaned houses for and not the community that I could be part of. I suppose it was an impostor syndrome that I was experiencing when I was 17, and it's still an impostor syndrome that sometimes I feel as a female founder."

Gomez started her career in technology as a contractor at Google. She was the Director of Operations for another start-up when she decided to take a pay cut to work as a contractor for Twitter, where she became a founding member of the international team and led Twitter's product expansion into 50 languages and dozens of countries.

"I felt that Twitter in its early years was pretty good at inclusion, regardless of the fact they weren't actually all that diverse culturally—there was diversity with regard to experiences and education, which I considered a big deal," says Gomez. "I remember joking with a VP that his Ivy League education meant nothing, he was hiring better product managers that didn't go to an Ivy League school, and he agreed. Anyway, not a lot of people were thinking about the diversity conversation in the same way we are today."

"I think the most magical years at Twitter were probably 2009 to early 2011," adds Gomez. "2010 was just awesome … it was the World Cup and we're like 'oh my God we have this platform that people want to use.' So yeah, I have nothing but good memories of my time there. As it grew into a big company, I think they were impacted by the factors that afflict most tech companies as they grow—the cultural factors, the entitlement, the pattern-matching in recruiting. It's part of Silicon Valley. But I have very fond memories of my early time there and the people I worked with."

Of today's tech industry, Gomez wants to know where the moral compass is pointing, especially as it relates to the displacement of communities in the Bay Area resulting from tech wealth. "Where are the opportunities for the working class?" she asks. "They're being displaced, their kids are being displaced, and these kids probably won't be able to access advanced classes or spend a lot of time with their parents because they're working all the time, and they're moving out to the East Bay. And it's not just the tech giants, it's the smaller tech companies, the unicorns, it's early-stage employees that got rich when their company went IPO ... it goes from an individual level all the way to corporate."

"I am speaking at a foundation gala that is raising money for college education. These may not be A+ students that end up going to the Ivy League, but they're good students that finish school and proceed to a four-year college. And Facebook is only a mile away. Why aren't there apprenticeships or internships for these kids? So I just think there has to be much more responsibility. It just kind of baffles me sometimes that we don't take into account any of this, the positive impact we can make as an industry."

Founding Atipica

Gomez wondered if the issues of diversity in tech that she has been seeing could be solved with technology itself. She started looking at how applicant databases function—analytics, data mining, and so on—and the more she talked to people, the more she realized the application process was shortchanging all facets of diversity. "Women don't assess themselves well, for instance, and they might not apply to a job because they think they aren't a perfect fit. So what happens? We don't apply, or we apply to the wrong job."

The goal of Atipica was to use big data to understand diversity in the recruiting pipelines. Along the way, Gomez and her team realized that all applicants, not just those other than straight white males, share common pain points in the application process. They talked to hundreds of HR and talent-acquisition professionals who told them about their frustrations as they tried to connect to the most compatible applicants without manually sourcing them.

Gomez was sure that technology could help. "Of the 75% of resumes submitted that are never seen by the human eye, most of those people don't have networks or referral systems. And they are most often the diverse applicants, whether racial, gender, background, educational, national, international ... they just apply through company websites and get lost. So that's how I started [Atipica]." She launched the company in January 2015, to provide optimization technologies for applicant databases.

Finding Support in Fundraising

Gomez credits much of her fundraising capabilities to a program called LaunchX, an organization that teaches female CEOs how to get access to institutional money. LaunchX was founded by Anna Khan, a Pakistani native and one of three female venture capitalists at Bessemer Venture Partners, the oldest fund in the United States. Khan noticed that, while more money is pouring into technology, fewer women have access to the dollars invested. Less than 3% of all venture capital raised in 2013 went to female-led companies, an abysmal number by any measure. While the number of women launching businesses is increasing drastically, the proportion of investments in female-led companies is not. Khan decided to create LaunchX, which offers women an immersive opportunity to learn how to raise capital for their companies.

"A guy can get away with Angel investment or some sort of investment in an idea even if it's not fully fleshed out. I think when you're a female it's harder," says Gomez. "But I've gone this far, I've had a lot of support, I've talked to really great partners who've given me excellent feedback. I pitched early, I pitched to the wrong friend, I pitched to the wrong partner, etc. And so I've designed my own strategy, and right now, that's where I'm at, on the investment side. I would never give up what I've learned over these past six months, just talking to investors. It's so fascinating."

Gomez notes that the lack of diversity on the venture-capital side is astonishing. "I think, for example, it's unlikely you're going to speak with a Latina venture capitalist on Santa Hill Road, or any immigrant woman venture capitalist. But I've learned not to look at the differences between me and the venture capitalist I am meeting with, because that's going to derail everything. Instead, I just focus on the similarities. Where did they go to school? What did they learn? Who are our mutual acquaintances?" Despite some obstacles and mistakes, Gomez currently feels very positive about raising money for female entrepreneurs, and is currently closing a round.

3.3 The Global Opportunity for Fintech Female Founders

"Be it the challenge of American women to achieve leadership positions or the struggles of Afghan women to even take on a job, the fight for access and equality is universal."

Targeting Untapped Markets

By Tahira Dosani, Managing Director, Accion Venture Lab

Tahira Dosani is the Managing Director of Accion Venture Lab, a seed-stage impact investing initiative focused on fintech and financial inclusion. She has also worked with LeapFrog Investments, the Aga Khan Fund for Economic Development, Roshan, Bain & Company, and The Bridgespan Group. Tahira holds an MBA from INSEAD, and BAs in International Relations, Computer Science, and Education from Brown University.

As a fintech (financial technology) investor, I work with entrepreneurs, male and female, who are driving innovation in the sector. I see start-ups grow and evolve, observing firsthand the implications of female founders and women in leadership and on boards.

For much of my career working at the intersection of technology and financial services, I've been the only woman in the room. Not the only woman in the organization by any means, but often the only one in the room where the decisions are made and usually the only person of color. It's hard, looking up and around and not see anyone like you. It's isolating.

We're All in it Together

Yes, we've made huge strides; in the United States today, women comprise half the labor force. But when it comes to leadership roles, the gender disparity rages; among the S&P 1500 companies, there are more CEOs named John than there are female CEOs.[1] I love this statistic—it elicits a much more powerful reaction than the mere fact that women comprise only four percent of those CEOs. Forget the comparison between men and women—we can't even win when it's a John-to-women comparison!

This gap is even wider in most emerging markets. Ten years ago, I moved to Afghanistan to work for Roshan, the leading telecommunications operator in the country. I joined the organization with the mandate of building a Corporate Strategy function, supporting Roshan's growth and ensuring continued market leadership. Afghanistan had almost no communications infrastructure until 2002, when the first GSM licenses were issued. If someone wanted to make a phone call, they had to travel thousands of miles and cross a border to do so. A lack of technology was not the only challenge Afghanistan faced at that time. Under Taliban rule, women hadn't fared well, being prevented from accessing education, let alone professional opportunities. In 2006, when I moved to Kabul, female participation in the labor force in Afghanistan was

1%. Roshan had about 1,100 employees, and remarkably, Afghan women made up 20% of its staff. Although primarily working in the company's call center, women could be found in every department within the organization. When I started, there were two women who worked on my team. As I reviewed their HR files, I noticed that one of them, despite having worked for the company for three years, had not taken a single day of vacation. Worried that she might burn out, I suggested that she take a few days off. She looked at me, horrified. "Please don't make me stay at home," she said. She then shared her story, telling me about the battles she waged on a daily basis to be allowed to leave her home and go to work. Although her own parents were progressive and wanted her to be financially independent and build skills and capabilities from working, her extended family thought that her leaving the home to work was a mark of shame. Each day, she fought for the opportunity to go to work. To her, a day off was not a welcome vacation or respite, but instead a day of confinement, of limitation, of succumbing.

That story is just one example of how Roshan made a concerted effort to empower women in an environment that was extremely challenging. Even so, every time I presented to the company's Management Board (essentially the C-suite of the business), I found myself in front of a table of men. I remember making my first big presentation to the Board, proposing a significant change in the company's pricing structure. When I asked a colleague who was in the room how he thought the presentation had gone, he responded by saying that he thought my outfit looked great and that I should wear that color more often—*no mention of my work*. Although I was living and working in an active war zone, my biggest battles were to be respected and treated as an equal.

Be it the challenge of American women to achieve leadership positions or the struggles of Afghan women to even take on a job, the fight for access and equality is universal. At first glance, these challenges look very different, but the essence of them is the same. The fight is borderless; it extends across culture, race, and class. Often, when we are the only women in the room, we feel isolated, but women around the world are sharing our experiences. Our battlefields vary, but we are fighting the same war.

One Is Not Enough

A few years later, at another organization, I was the most senior woman and the only woman of color at the time when I joined. The CEO requested that my picture be placed at the top of the team section on the website to highlight the "diversity" within the organization.

A token woman is not diversity. Commitment to diversity and gender equality doesn't stop at one hire or one brown female face on a website. And recognizing someone *because* she is a woman can undermine her achievement. It is not enough to bring one woman in through the door and declare victory. Diversity is not a box that needs to be checked off. Breaking the glass ceiling is undoubtedly important, but to actually succeed we need to place a *constant* emphasis on achieving diversity of perspective within organizations.

Technology as a Driver of Empowerment

As we think about the evolving role of women in driving the creation and spread of technology, we have to consider women as consumers, as employees, and as founders and leaders. In each of these capacities, women need to be financially healthy. And to make effective personal and professional decisions as users, managers, or entrepreneurs, it's critical that women have the right financial tools.

Technology is supposed to be an equalizer, and it can be. I saw how simply owning a mobile phone enabled Afghan women to be reachable, to build networks outside their homes, to build independent lives. At that time in Afghanistan, less than 3% of the population had a bank account. Women made up less than 0.2% of account holders in the country. I worked with Roshan to launch M-Paisa, a mobile money platform that would enable anyone with a mobile phone to have a mobile wallet, enabling a level of financial inclusion and independence that was previously unfathomable for the population, particularly for women.

I currently lead Accion Venture Lab, a seed-stage impact investment fund that focuses on fintech and financial inclusion. We provide capital and support to early-stage start-ups that leverage innovation in order to improve access to or the quality of financial services globally. We look at hundreds of innovative start-ups each year, and that gives me a unique perspective on what it takes to start and scale a successful technology business. I also work closely with our founders and management teams, thus helping our companies grow.

Like the M-Paisa example above, many of our portfolio companies use technology to create, deliver, or reduce the price of financial tools. Our investees are at the cutting edge of mobile payments, better ways to borrow and save, lower-cost remittances, alternative credit scores that use behavioral and psychometric data instead of relying on a formal financial history, and other innovations that enable the underserved to build their financial health.

Only one of the 24 companies we've invested in to date has a female CEO (that 4% puts us right in line with S&P 1500 CEOs). However, five have female founders or co-founders, and the majority have women in key roles in senior management. Incidentally, the Venture Lab team is made of more women than men, a rarity in the venture-capital world. Today, our portfolio is still small and our companies are still young. I can't draw direct correlations from the presence and number of women in these leadership roles to the performance of these businesses, but if women make up half the target consumers of these businesses, it stands to reason that having women on the teams will help the design, development, and delivery of products and services for female consumers. Research has also shown the value of women in leadership and on boards for decision-making and growth in businesses.

So What?

Women need to actively and consciously create opportunities for other women. More female investors should mean more funding for female entrepreneurs. More female entrepreneurs should mean more women on teams. More women in high-growth businesses should make it easier for other women to enter the workforce. It's women supporting other women that will push us beyond the tokenization of women in organizations.

This is not to absolve men of responsibility here. It is critical for men to be aware of and understand gender issues in business and technology, and to be committed to change. Because men (even those not named John) typically hold leadership positions in companies today, they have the ability to set a culture, to introduce policies that better enable organizations to recruit and retain women, and to create a more inclusive environment.

Reference

[1] http://www.nytimes.com/2015/03/03/upshot/fewer-women-run-big-companies-than-men-named-john.html

4

Mentoring Investors and Entrepreneurs

4.1 Supporting Female Entrepreneurs and Creating Role Models in Israel and Beyond

"It's very obvious that you won't have enough female entrepreneurs when you have only male investors."

Moran Bar is the Co-Founder and CEO of GeekMedia, the parent company of Geektime.com. A seasoned entrepreneur, her high-tech experience includes an extensive background in storage, communications and social media. Bar participated in large, web-based projects such as 2eat. She is a certified attorney with specialties in private law and technology.

Geektime, based in Tel Aviv, is the largest international tech blog located outside the United States. It focuses on global innovation and on highlighting start-ups from across the world. In addition to their in-house content, Geektime's start-up contributor network provides a voice to start-ups based outside the United States, which comprise some 80% of the global start-up market. Geektime thus helps investors, entrepreneurs, fellow geeks, and tech enthusiasts cut through regional boundaries and compete in the global marketplace. Geektime boasts a reader base of more than a million unique visitors a month, and is also a partner of and contributor to many of Israel's leading tech events, providing media coverage of the latest and greatest coming out of the Israeli start-up, IT, and entrepreneurship scenes.

Rahilla Zafar conducted the following interview with her.

What are the key obstacles that women in tech must overcome?

I think there's a huge problem here of women staying in their comfort zone. If we have women-only meetings, or platforms of women only, this works great, but when you put them with men, they lose all their confidence. I hate it.

This is why I'm giving tickets for free to our events. Take our Geektime Conference, which happens once per year. It's an amazing competition of ten start-ups presenting ideas for the first time, on stage. I had to write a special post calling for women entrepreneurs—only then did they apply to the competition.

When we have more women writing about women's products or women's companies, that's when more women will get funded. And after that, we'll have more women investors. In Israel, I can count the number of female investors on two hands. Someone from Silicon Valley wrote something very interesting—how do you expect a woman's company, for which the consumer base is women, to be funded when the investors are all men? They need to go to their wives and ask if it's worth the investment.

We have the same problem here in Israel, and I don't think that we've made any progress yet. It's very obvious that you won't have enough female entrepreneurs when you have only male investors.

When did you first become interested in hi-tech?

In high school I studied biotechnology. I didn't learn anything about computers. I don't come from that background. And I'm a lawyer. But in the army, I served in a computer unit. It was at just the right time, in 1997. It was part of the first bubble of the nineties.

And since then it's been great. I've made lots of money. I used to do integration, IT, security, and stuff like that. I've worked mainly with men, but I'm not what you call shy. Most of the time they thought I was the secretary, not the one who's going to do the job—in the IT industry, it's very rare to find women. I embraced this and never backed down.

How did Geektime get started?

Seven years ago, I met my partner for Geektime. At the time, no one was writing about the start-up scene in Israel. They all wrote about financial stuff, the big companies, the big enterprises. No one wrote about funding the start-ups themselves. So we started doing it as a hobby.

Today, we have thirteen people on staff. We've raised money from investors, and we are now expanding globally. We are going from ecosystem to ecosystem. We write about Finland, about Japan, about China ... It's not necessarily the usual suspects. I think that, eventually, the majority of innovation will be global and not from the United States.

Tech above politics

Geektime was the facilitator here in Israel of start-up weekends for three years, from 2010 to 2013. There were participants from the West Bank, Gaza, and all over Israel. Tech is politically agnostic. I don't care who's working with me. The only thing that I want to know is that he or she is the perfect person for the job.

The media loved it, but we pushed them back—we said no media—so there won't be any problems for attendees. We just wanted to work, and that's it. I believe that tech is above any politics. You're not being judged for who you are but by what you can do. Tech can create a lot of bridges and break down a lot of walls.

What do you think entrepreneurs need in order to succeed?

I think success is mostly about character. My parents told me that I can do anything. This is where it starts. Parents need to say, "Just try it. Maybe you'll fail, and we have your back if you do." This is how it works. I'm also married to a woman who supported me for a year and a half when we started Geektime. So it's all about support from home.

For me, I think that raising money was hard (because here in Israel, to raise money for media is almost impossible), as was building products, finding the right team to build them, and finding the right company culture for this team. Company culture—that one took me a lot of time. Hiring the wrong people is a mistake that everyone makes, and you need to learn from it.

Personal connections have been a huge part of our success. We've been in touch with investors long before asking them for money. We created the relationships. Asking for an investment, it's not something that you will get by saying, "Hi, my name is X. Give me money." It doesn't happen like that. You need them to trust you and get to know you.

What has been the biggest challenge in running Geektime as a female CEO/co-founder?

Monetizing media is very hard. The models are changing. Advertising is not advertising anymore. The industry is shifting. Not only that, but I had to prove myself again and again in an environment that was made up mainly of men. But if you stop, you see that the fear is in your head. And if you just ignore it, you can just be yourself and present your product, and that's it. We need to break the cycle of fear, and we can do that through support, I think, and by setting examples.

Women need to see, in the media, that there are many other entrepreneurs, female entrepreneurs, who are willing to help. We need more women on stage and on panels—and, of course, we need more women investors. That's the key.

How do you deal with issues such as maternity leave and salaries?

We have to return to the main problem of salary inequity and maternity leave, and the fact that when you give birth, you stay at home and not your spouse because he makes more money. That's a lose-lose. The moment that wages are the same between women and men, I think things will start changing.

The other obvious thing is that we should have more women politicians. More women in front, more women managing companies, more women on boards, more women—this is how we'll do it. It won't happen until we will push it.

When I'm interviewing here at Geektime for positions, a woman can speak highly about herself when it comes to product or performance, but when I ask about salary, she acts completely differently and says something such as, Ooh, you know, I think that I should get X, and I don't know, maybe I'm not worth it. This is what happens every time and I yell at them for it!

In terms of women in the field, while we have an equal chance to get here, you need to go back to the fact that in high school, fewer women are studying STEM-related subjects. You need to go back even to kindergarten, where little girls are encouraged to play house and boys are going to build stuff. It starts there. I say it all the time, and I really believe it.

Today there are schools and other environments who'll give the same opportunity to boys and girls to experience everything equally. That's one thing we're doing right. But it's not enough.

So, I volunteered in an initiative called The Cracking of the Glass Ceiling. We went to a group of 13- and 14-year-old girls and asked them, "What do you want to be when you grow up?" They all said, "I want to get married. I want to have children." That was, like, the main goal of their lives. Then we built websites with them and talked about women role models, such as scientists. Their perspective expanded, and they all learned that there are women role models to follow. That was a great program for girls. There are similar programs everywhere that are encouraging more girls to go and learn math or computer science.

4.2 Female Entrepreneurs Making the Connection

"I struggle with the question, 'Why aren't there more women in tech?' That's too simplistic, and we're not moving the needle by constantly asking that question. We're just cementing the view that there's a problem."

Aurore Belfrage started her career as a scrap dealer in Stena Metall Group, where she was managing director for the company's Czech and Italian scrapyards. She recently joined Stockholm-based EQT Ventures, where she heads up the matchmaking platform Together for early-stage start-ups and angel investors. She co-founded the hyped global gift-card start-up Wrapp, in which Skype founder Zennström and LinkedIn founder Reid Hoffman were early investors. In 2014 and 2015, she partnered with the current-affairs site Your Middle East and hosted a successful Middle East road show about start-ups and entrepreneurship, "Saluting the Crazy and Naïve." She also manages a portfolio of Middle Eastern start-ups and investor clients at Busy B Ventures.

What follows is an edited transcript of an interview conducted by Rahilla Zafar.

How did you make the transition from being a scrap dealer to working in the tech sector?

I worked for several years in waste management, scrap dealing and recycled electronic waste. I was a country manager in Czechia and Italy. By chance, in 2008, I was offered a job in the tech and digital space. My first reaction to the opportunity was, "It's impossible! I have no clue what you do! I'm a scrap dealer!" The person recruiting me explained, "You'll learn it in two weeks." He needed to rebuild a product offering, and I had the "scrappy" skills to do it.

So I moved from waste management into digital in 2008—and he was right. I learned the space, both the technical opportunity and the magnitude of the communications aspects. This was in London. When I moved back to Sweden, I joined the global online product team for Metro International, a global print publication that needed a digital transformation.

In 2011, I was invited to co-found Wrapp, a mobile social-gifting company. It was a start-up journey that was classic in the sense that we were first super-hyped and then not so much. We raked in a lot of investments early on.

The founder of Skype was one of our investors, and then the founder of LinkedIn came on board. We made so many strategic and tactical mistakes on our journey, and one of them was to move to Silicon Valley for all the wrong reasons—in my opinion because of vanity. It was a very steep, very interesting learning curve in start-up land. I loved it.

You've been very inclusive of women at the events you organize, but have not advertised this. Why?

I struggle with the question, "Why aren't there more women in tech?" That's too simplistic, and we're not moving the needle by constantly asking that question. We're just cementing the view that there's a problem. During my events series, in every city I visited in the Middle East the entrepreneurs I brought on stage always included women. I had to deal with a lot from many of my sponsors because they wanted to blast it in the media that they supported female entrepreneurs. I said, "You're not allowed to gain short-term PR wins and look good as you are just cementing the view that there's a problem. Be happy to know that you put amazing entrepreneurs that happen to be women on stage." That, to me, is much more powerful.

I was reading an article about how to raise seed money, and the article validates my point of view, which is that raising money requires emotional intelligence and emotional connections. Early-stage and angel investment decisions are made mainly as gut decisions and then validated by Excel. You, an investor, if at first you like *me* and you will want *me* to succeed, then we will use Excel sheets to validate the business case and work together. Either way, I'm triggering some sort of emotion in *you*. And I'm a firm believer in this. I've worked in sales all my life. My theory is that women are less successful in this phase because we try hard *not* to show any emotion. Why? Because we are pre-stamped as emotional, and we're worried that if you like me it's because there's some flirting going on. So instead, we tend to be unnecessarily square and factual, and investors don't connect on a human level, and therefore we find it harder to get the investment. Business is done best when trust and genuine connections are formed, and it has nothing to do with flirtation, so women shouldn't be reticent about building strong relationships with potential investors, male and female.

Also, when I'd moderate a mixed panel, I'd do some experiments. I would ask the women founders to tell me about their start-up, the capital they raised, and the records they broke—all business. But when I interviewed a man, I'd start with the same questions but then, in the middle, ask, "So how do you manage work-life balance?" I'd stay completely straight faced, and he'd start laughing. The audience would start laughing. Everyone understands that this question has never been asked to him, but always to women, and then he has to answer. Everyone in the room got the absurdity of the question and the gender issue without me having to explain a thing. I think it's a more elegant way to make the point.

4.3 The Importance of Finding "Champions" for Entrepreneurs

"While there obviously need to be more female investors, such as Joanne [Wilson], we also need men that invest in, believe in, and coach women."

Carla Holtze is the CEO and co-founder of Parrable, an innovative technology platform that provides third-party digital-identification solutions to the data and marketing technology ecosystem. Prior to Parrable, Carla worked at the BBC, The Economist Intelligence Unit, and Lehman Brothers in both New York and Hong Kong. Carla earned her MBA from Columbia Business School, her MS in Journalism from Columbia University, and her BS from Northwestern University. Carla serves on the Advisory Board of the San Francisco Symphony Soundbox, Symphonix, and serves as a mentor for emerging entrepreneurs and technology companies through Start-up Mexico (SUM).

What follows is an edited transcript of an interview conducted by Rahilla Zafar.

How did you decide to start your own company?

My story with entrepreneurship started a long time ago. I grew up in Des Moines, Iowa, and from a very early age I always had a knack for dreaming up businesses and taking new products to market. When I was in first grade, I made butter in my kitchen and sold it out of a red wagon, taking it door to door. I outsourced my mom for production. My closest friend, Allison, and I also made "firestarters"—wrapping sticks we found in the woods with some gluey concoction we whipped up in the bathtub and laid them in the driveway to dry. Oh the sight. We also sold those "firestarters" door to door to our neighbors who had grown accustomed to supporting our entrepreneurial prowess. I always had a twist on your typical lemonade stand, and from a very early age I wasn't shy about taking some crazy product to market. I was very much encouraged by my parents to take the moonshots and not to be afraid of the stumbles or of failure. I know that it has very much influenced who I am today.

When I was a child, one of my uncles started a computer company that ended up going public. It was a huge inspiration to watch that company grow from a start-up on a cattle ranch into a successful, global company. I saw the effort, tenacity and team effort that it took to build a company, and my eyes were forever opened to the world of technology and the Internet.

What did you do after college?

At Northwestern University, I completed a dual degree in economics and communications, as I was fascinated with media and the business around it. I didn't know what to do after college, so I went to Wall Street. It wasn't my passion, but I learned a lot of great skills, and probably most importantly, what working in a large organization is like. I gleaned many lessons that are crucial for any person who wants to ultimately run a big company. I did that for four-and-a-half years, and I grew throughout the organization, living in both New York and Hong Kong. I had always aspired to start my own business, and I knew that the global experience would teach me invaluable lessons that I could carry forward.

In 2007, I was ready for the next challenge, and I went to graduate school. I put together my own dual degree at Columbia University, I got my masters in Journalism and my MBA. I was really interested in the media world, and that was at a time when Twitter was just coming out. I saw that the world was going to change, and especially with mobile. I was really interested in initially helping media better monetize as the world was going from print to desktop to mobile.

When did you start your first business?

After graduating from business school, I started a company that was initially trying to do personalization for e-commerce companies. We were trying to re-envision what the world meant, such as with personalization. We believed that everything online and on mobile should be personalized for and relevant to the user. When you pick up your mobile phone, everything that you see, the content, the experiences, the advertising, everything should be personalized and relevant to you.

Flashback to our old business, where we had these interactive, gamified rich media units living on publisher sites that were streaming in products from e-commerce companies. What we were doing was extracting user preferences by asking them to swipe right and left, and asking them what they liked. The real gold was the data and the intent-based profiles we were creating. We wanted to use that data to help users discover new products that they would like and to give them one-to-one offers on mobile. It was all about offering personalization to optimize mobile buying experiences.

We had really great clients and partners that were signing up to do case studies with us to show our 47%+ engagement rates. However, we ran into a huge problem on mobile that was limiting our scalability.

Fragmentation on mobile prevented all the great data that we were collecting from actually being usable. On mobile, there are third-party cookie restrictions in Safari, which represents approximately 75% of the United States mobile web market. Moreover there's the app world and the mobile web world, which don't connect. Then add in the web views, which happen when users click links outside an app. Unfortunately, on mobile, the dots don't connect, and a user looks like millions of different people on the same device. And if the dots don't connect, the data doesn't connect.

That means that if you're on an iPhone and browsing around on a shopping website and then you jump over to a news site or an app, for instance, an e-commerce company will most likely not know who you are. It's hard for e-commerce companies to reach the right person and drive conversions, and it's hard for the media world to monetize that anonymous audience. It was a massive problem.

We drew out the problem every day on the whiteboard in our office in the Flatiron neighborhood of Manhattan. Then we stared at it the next day. We went and we tried to license the solution from some really big companies, and they were like, "No. We don't have the solution." We didn't take no for an answer and ended up engineering a solution. That solution is what our business became today.

How did you end up pivoting your model?

We still had the old media business model at the time, and about five months later, one of my tech advisors, Justin Greene, who had been my advisor through our old business, said, "Carla, why don't we just can the old media business? That's not the scalable business. Why don't we focus on Parrable as the identification platform? The whole world needs this solution. It needs a third party to do this." He said, "I will come on as your CTO and co-founder to do this."

That was one of the biggest, exciting moments, where I could have an experienced technical friend who is also a CTO, who had built and sold many businesses, as my co-founder to help build out this business. Parrable is a third-party identification solution and what we're doing is unifying the user's profile across the entire device. We are deterministic. We don't use fingerprinting or probabilistic methodologies, and it's privacy-safe, which is very important.

What does this mean in the grand scheme of things? Right now, Facebook and Google have a duopoly. They are minting money because you log into those platforms and they know who you are. They collect the data and the dots connect in their respective walled gardens.

If they know who you are and what you like, they can follow personalized, one-to-one experiences. For example, if you're a pizza company trying to show pizza lovers relevant offers or photos to make their mouths water—you're going to choose to do it inside one of those walled gardens, such as Facebook, that can guarantee that you will indeed be reaching a pizza lover. And then, they will show you how successful you were in reaching that target audience.

This is precisely why those companies are doing so well—they know who the user is, they have all the data within their walled gardens, and they generally own the media.

For the rest of the world that's outside the walled gardens of Facebook and Google, it's a real challenge. Media companies can be challenged to monetize effectively on mobile, and marketers are challenged to reach their addressable audience. Parrable's third-party identification solution provides that identification platform that levels the playing field for the rest of the world. We're trying to democratize the data and marketing world, so that everyone can play.

What's your advice to women who are looking to raise money?

Women often don't speak as forcefully. They wouldn't say, "This is the deadline. Are you in? Are you out?" We're not tough like that. I've had to work on it. One thing that we have to work on, and that a coach can help with, is being a lot stronger in the way we speak as women, thus showing that we have the big vision and dream. Being comfortable speaking about it in a forceful way and having conviction in yourself and your team may not be natural.

It's not about lacking self-confidence. I try to communicate effectively to male investors but I know it requires a conscious effort to do it in a way that resonates with them. Training and coaching women on communicating are critical if we are to build businesses that make a lasting impact.

Joanne Wilson is one of your investors and one of the most prominent supporters of female entrepreneurs in particular. What has it been like to work with her?

Joanne has been a huge advocate and is extraordinarily supportive. She's tough in a good way, and thoughtful. She really stood by us even when our business changed a lot, and it's gotten a thousand times smarter. The value, in my opinion, has gone up enormously, and she's been there supporting me. Investing in an entrepreneur is almost like raising a child.

We need more women such as her to help coach women and get them to create the technology companies that can IPO. It's like the saying, "It takes a village to raise a child." It's the same thing for an entrepreneur. It takes a whole community to support you, and she's been great.

While there obviously needs to be more female investors such as Joanne, we also need men that invest in, believe in, and coach women. Male champions are important. Mike Lazerow and Ken Arnold are two investors of mine that have been incredibly helpful—they have taught me how to communicate more effectively to men, have advocated for me, and introduced me to their communities. Both have been incredibly successful, and both are making huge contributions to the entrepreneurial ecosystem by believing in and helping to coach women such as myself.

What are some of the things you've done personally to help support women in technology?

What I'm really passionate about is empowering women. We have an office in Mexico and have two extremely talented women working there in engineering and business development.

I try to give some really great opportunities to women, whether in Mexico or other parts of the world, where it's really great for them to be able to come into positions of leadership and take on more and more responsibility. So they can later turn around in their own communities and mentor women and men. They are becoming role models and experts.

4.4 Bringing Water to the Women's Capital Desert

By Tracy Killoren Chadwell – Partner and Founder, 1843 Capital

It was the year 1843 that Ada Lovelace, the only legitimate daughter of Lord Byron, wrote the first computer program. Had she been taken more seriously, computers would have been developed 100 years earlier. In Victorian England, women were expected to needlepoint, play the piano and engage in a charming manner. A woman fascinated by algorithms was considered unusual and slightly dangerous! Women innovators in technology today continue to suffer from unconscious bias. This bias is especially strong when it comes to venture capital funding. Women received about 3% of total venture capital dollars from 2011–2013[1], while they account for 18% of start-ups.[2] This is in stark contrast to all the data that suggested women founders do more with less capital and their companies are more profitable.[3] I founded 1843 Capital, with my Partner Maria Hancock, to address this funding desert for women. 1843 Capital makes early-stage venture capital investments in women-led technology companies. By providing critically needed capital at this stage, the companies receive enough life blood to prove their concept and make the benchmarks needed to secure the next stage of capital or become cash flow positive (meaning they actually make money).

Helping women with technology companies doesn't begin and end with capital. It is a critical catalyst for success, but by no means the only thing. When I invest in a company I help them:

1. Recruit board members,
2. Think about their marketing strategy,
3. Recruit team members,
4. Find clients,
5. And most importantly, prepare them for the next round of financing.

Fortunately, there is an entire ecosystem devoted to helping female entrepreneurs start companies. Universities have been terrific in supporting all entrepreneurs and some have taken the next step to support females specifically. Some great examples are:

- **Cornell University**: They are building a women's technology community by organizing the Johnson Women in Tech day.
- **The state of Connecticut** has a program called Business Express. For organizational help, there are accelerator programs for women-owned

businesses. Springboard, Mergerlane, Echoing Green, and the Refinery are a few.
- For **seed funding, crowdfunding** sites like Portfolia and Plum Alley are welcome additions to the traditional sites.
- The **SBA** provides **low-interest loans** to women.
- **Private lenders** that are focusing on women: such as Credibility Capital and Bond Street.
- **Seed stage investments**: Golden Seeds is an angel group focused on women founders, as is Plum Alley and Astia.
- **Accessing capital**: The Vinetta Project and Global InvestHer can assist with accessing capital.
- **Seed stage venture** capital funds that are willing to invest early: Jump Fund, Women's VC fund, and Built By Girls.

There is a tremendous amount of support for women businesses at the early stage. However, as these companies grow, there are fewer options for funding and that is the gap 1843 Capital aims to fill.

My path to venture capital wasn't a straight line . . . more like a spiral staircase. I was always going up, but not always in the same direction! Like most women, I had many strange and sometimes illegal experiences. More than once I discovered a man in my position was making more than twice what I made. I was left out of "meetings" that happened at strip clubs, and had one boss take off all his clothes to change on his jet. Early on, one of my bosses asked me to come with him while he went to look at the prostitutes in Hawaii. The most frustrating though is the unconscious bias that comes from other women, and and I still feel it every day. My secret weapon is that I use "being underestimated" as fuel for growth and accomplishment.

At 12 years old, I learned to code in BASIC. I loved it and found myself doing the homework for everyone in my class. But that was it. We only had one introductory class. I didn't have access to a computer, so I would take the vacuum and televisions apart to satisfy my engineering thirst. I was definitely "tech support" for the limited technology we had in the house (Betamax, Sony Walkman, and the Atari Pong game). I thought I wanted to go to MIT. My mother talked me out of it saying "the boys aren't cute enough." So instead of engineering, I ended up going to law school like my dad.

My first job as a lawyer was as a receptionist. When I graduated from law school in 1992 the economy was still terrible. Even though I had done all my homework, differentiated myself and been on tons of interviews, I couldn't get an offer. I went back to the drawing board and did more homework. Going for an associate position in a law firm was clearly not working. Because I was reading all the trade publications I could get my hands on, I saw an article about a new merchant banking firm opening in Chicago, in Crain's Chicago Business. The firm SCM International, was to be headed by Senator Adlai Stevenson, who was the former head of the Senate Foreign Relations Committee. I cold-called them asking if I could come in and discuss working for them. They said they weren't hiring attorneys right now so I asked if they needed anything else, and their response was that they needed a receptionist. I responded that I would love to interview for it. It was a way in the door. I had no interest in being a receptionist, in fact it was a big blow to my ego and expectations. However, it gave me an opportunity to show how good I was, in an organization that completely fit my skill-set.

When it came time to negotiate the position and the salary, I was not in a position of power. After all I had been through, I felt really lucky to get the job. I accepted the salary they offered me, although it was less than a third of what my fellow classmates were making in law firms. But I had a start. I was in an industry that was interesting and an area that was growing. And on top of it, I negotiated with them that I would be their receptionist for six months until they found someone else. At that point I would move into an office and help with legal and finance work. After six months of telephones ringing in my ear, I moved into an office and started working on spreadsheets, term sheets, and corporate formation. A couple years later, I helped buy Japanese-owned office buildings for Sam Zell. It was a great experience but I was starting to feel a glass ceiling in Chicago.

Time to move to New York. I broke out my book of business cards (no LinkedIn at that point) and called everyone I knew. I asked if they knew anyone who worked in private equity. Zell had both real estate funds and private equity funds investing in companies. I had become very familiar with the way funds were structured (limited partnerships) and run and wanted to be a part of it. I didn't get very far with private equity, but a friend did introduce me to someone at Robertson Stephens. This was an investment bank headquartered in San Francisco and knee deep in the first wave of the technical revolution. This technical revolution was fueled by a relatively new thing called venture capital. Venture capital is money provided by investors to start-up companies with perceived long-term growth potential. The investment is made in the form of equity (stock). This is a very important source of funding for start-ups that do not have access to normal equity capital markets or debt financing. In other words, the investments were so risky, no one else would give them

4.4 Bringing Water to the Women's Capital Desert

money. Most venture capital investments fail . . . but the ones that succeed do so spectacularly. By pooling together multiple investments in a venture capital fund, the risk is diversified and the blended return of the winners and the losers can be impressive.

The capital in San Francisco during this time flowed in a circular fashion (much like today). People would start companies. The companies would get bought or go public (sometimes three or four companies went public in one day at that time), leaving the founder with a large amount of cash. The cash would get invested into venture capital funds, which in turn would be invested in new founders and new companies. It was capital recycling and all the capital and expertise was invested in men. I was hired in the part of Robertson Stephens that focused on helping venture capital funds get started. I helped structure their funds, position their companies, and connect them with investors. I already knew about fund formation and how companies were financed, but I had to learn about networks, circuit boards, semi-conductors and software! I was in heaven.

As a result, I joined Baker Capital in 1999 as one of the only female partners of a growth capital venture capital fund. I think in New York there were four or five of us.

John Baker started his fund to take advantage of the opportunities in the communications space. He had formerly been the head of telecom investing for the firm that had started venture capital investing in New York, Apax Partners. The venture capital community was small in New York at the time and I had met John while I was working at Robertson Stephens. It was the early days of the next wave of technology innovations. The companies we saw were the early versions of cloud data storage, voice recognition software (it hardly worked then, due to individual pronunciation styles), and technology that used sensors to gather data. We all use the app WAZE today to figure out the best route to get home, but in 2000, it was our investment in Traffic.com that led the way. There was no "crowdsourced" data or GPS on phones, so Traffic.com used sensors on the highway to count cars and give estimates of travel time.

In 2003 I had two children and decided with the economic turmoil it was a good time to take a break. It didn't last long, and I couldn't stand still during that time. I oversaw renovating our house (never, ever, do this), coached teams at my son's school, and chaired a fundraiser.

About four years ago, it occurred to me that all of my investments were managed by men. I had seen a lot of data suggesting that women were better managers of companies than men. I decided to take a portion of my capital and invest directly into companies run by women. It took a while for me to get into the flow of potential deals.

However, I knew when I met Gregg Renfrew of Beauty Counter that she was doing something very special. Beauty Counter was the first beauty brand to make and distribute products completely free of over 1,000 chemicals banned in the EU. In the United States, we ban four. She combined her "never list" formulations with gorgeous packaging and a unique distribution model that uses sales consultants at home, rather than selling to retailers who take a much larger percentage of the profits and may not market or position the products well. Her company has been an incredible success, growing at more than 300% per year. Gregg was one of the lucky ones. She had, on a relative basis, an easy time raising capital. She was a second-time entrepreneur who had successfully sold a business. She also had a long-term and solid network. However, younger women who have less established networks have a harder time raising capital. I have seen for myself how women with strong ideas and great momentum in their companies have failed to raise money. I have also seen young men raise millions of dollars on an idea and bravado. This was recently backed up by a study done by Wharton, which found that men overall were 60% more likely to pitch successfully to investors than women, even when the content of the pitches were the same.[4] There are many factors that go into this, but one of the largest reasons is the cyclical nature of venture capital. Men have historically founded, sold, and funded new companies. When I worked at Robertson Stephens, I never encountered a woman-led company.

Inspired by the success of Beauty Counter and other female led companies in my portfolio, I decided this year to form my own venture capital fund. I decided to form the fund in the New York area because New York is a booming center of tech innovation and has always been a leader in financial services, media and retail areas. It is no surprise we are seeing incredible women-led deals in the area of fintech, adtech and ecommerce coming out of New York. Purely from a volume perspective, New York is leading the way. A recent study by Crunchbase showed from 2009–2014, New York City had 374 companies with a female founder that had received funding. That surpasses 338 during the same period in San Francisco.[5] With all the new companies and all the new funding sources, we are getting closer to getting badly needed capital into the hands of women, but not the entire answer. It is my hope that more successful women-led companies will encourage more investment by the private and the public sector into this efficient and fast growing sector of the economy. I hope soon my "investment thesis" will no longer be relevant. With 1843 Capital, we will see a dent in both the investor and the founder statistics, and change the way the world sees work.

4.5 Think Big: Expanding Plus-Sized Fashion Globally

By Mitali Rakhit, Founder and CEO at Globelist

Mitali Rakhit is the Founder and CEO of Globelist and STUNNER. Both companies seek to provide access to fashion for curvy women in the Middle East through B2B and B2C channels respectively. She has earned an MPH from Yale University and writes for several publications such as Fortune, Forbes, and Tech in Asia. Mitali is passionate about women's issues and economic empowerment.

My journey into the world of technology start-ups actually began after multiple rejections from the world of consulting. I would almost say that it was "accidental" in the sense that, although I always had entrepreneurial ambitions, I never thought they would be realized until a much later stage in my life. Most of what held me back prior to when I got started was fear and a lack of adequate preparation. It was only a few months into the process that I realized that you never really become completely fearless or totally prepared, and that accepting this fact is one of the first signals of your transformation into a true entrepreneur.

When you are a woman of color in the start-up world, you quickly realize that you have multiple forces constantly threatened by your presence and pushing against your progress. It was at one of my first networking events in New York that I had my first realization of this in a big way. A professional South Asian man in his mid-30s came up to me and asked what I was working on. At the time, I had envisioned my concept to be an e-commerce store for South Asian designer wear. He laughed at me and asked how "a tiny average-looking Indian girl expected to be able to compete with six-foot-tall Jewish guys" for funding. At the time, I was just in shock. I didn't know how to respond, and I walked away in silence. I was never so aware of my physicality or its importance, but it then seemed that others might be. The encounter did produce some positive outcomes, however. It forced me to rethink the size of my target audience and aim for an idea that was much bigger than I had initially started out with.

Armed with a larger vision, I then set out to Singapore for Fashion Week. While there, I spoke with many designers to try to recruit them onto our platform. I passed by an Indonesian swimwear duo who saw me ogling their Black Forest cake and offered me a place to sit and have a bite. I eagerly accepted and began a conversation with them about New York City, and that eventually led them to provide me with a candid review of the revised product idea and a lengthy explanation of what was actually needed: distributors. They told me that what I was doing was OK, but that, if I instead focused on a B2B solution, I would definitely get much more traction and interest. So, I went back to my hotel room, did a quick run of the financials of the new model versus what we currently had settled on, and decided to take their advice. This meant throwing away an MVP that I had already spent several thousand dollars on developing, and starting from scratch with a new CTO that I had just hired based on one meeting and an introduction via CoFoundersLab. Sounds risky now that I look back on it, but it was the best decision I ever made.

The next challenge was refining the B2B model and assessing the strengths and weaknesses of products that were already available on the market. Unlike the B2C arena, there are no clear winners in B2B (AliBaba had not IPO-ed at the time and was not yet a media sensation). We had to figure out why the current solutions weren't working and how we could make them better. After much research, we focused on a few key points. The first was that the most significant hurdle to all of the companies then developing solutions to the problem was the inability to attract buyers. If no one is buying what you are selling, you do not have a business. The second was that, in order to attract buyers, you needed to make something available to them that they did not currently have access to. In the current landscape, other companies were trying to do that by introducing new technology and asking vendors and retailers to adopt it in the hope that it would make their lives easier. However, the technology had a significant start-up cost, and that meant a big risk for initial customers. The other problem had to do with continuing to use the technology even when customers ran into bugs and glitches. It was hard to get anything to stick, especially when thinking of technology in the apparel industry, which is, by all accounts, archaic.

Expanding in All the Right Places

We had to identify a product category that would be attractive to buyers in a market segment that was growing rapidly and required help for distribution. It was then that plus-size clothing came onto the radar. It is a nascent market globally, but more developed in the West than anywhere else. According to CNBC, the plus-size market in the United States is worth $17.5 billion a year, but only comprises about 18% of total apparel sales. I started out by contacting buyers domestically, but most vendors had a wide domestic network and wanted to focus more on growing their international business where they had fewer contacts and connections to win sales. Unfortunately, Singapore and South Asia were still not yet ready to justify a significant investment in the plus-size segment. In addition, a culture centered on shopping as a social activity, high spending power, and a passion for fashion make the environment ideal for the plus-size industry. In fact, the plus-size opportunity in this region could reach up to $10.3 billion if the apparel purchasing patterns continue. Everything started falling together until I realized that I would probably need to move to the other side of the world, away from family and friends, to turn this dream into a reality.

My first trip to Dubai was great. I met with several kind and welcoming people in the fashion industry who were eager to hear about new ideas and disruptive technologies. It became apparent that there was a small but thriving start-up culture in the region that was hungry to be recognized on a global level for its contributions, and that welcomed entrepreneurs and investors who would help to bring those efforts into the limelight. There was not much that I observed happening in the fashion-technology space other than e-commerce, which indicated the presence of a huge opportunity. The MENA region's focus, and its large investment in its retail sectors, also looked promising in terms of growth for the next decade. The macro-economic climate was aligned with our goals, and the opening of our Dubai office appeared imminent. Being physically located in the region would be critical to our success, because business relationships in the region are developed face to face.

As an Indian American doing business in the Gulf, I am often asked many questions about how I was able to identify and enter the market. I am sure as the years go by we will be asked how we are able to survive and maintain a competitive advantage. The answer I give is the same one that any entrepreneur in New York City would give. We researched the market and found the best macro-economic conditions that would develop with or without our company's existence. We then began reaching out to our networks for help in assessing market-entry strategies, and picking the one that made sense to us. Since doing business in the Middle East necessitates a lot of face time and human interaction, opening up an office there and maintaining a significant local presence were critical. Because I am a woman, many people assume that the Middle East would be a less ideal place for me to do business in, especially compared with the West. I think their assumptions couldn't be more flawed. The region is well known for having almost three times more women entrepreneurs than the United States. Personally, I have not felt a greater degree of discrimination in the region than I have in the United States—on the contrary. Many male friends who are Gulf Cooperation Council nationals have strongly encouraged me to expand our business to Saudi Arabia, because of the large market opportunity, while male friends from other parts of the world have frequently discouraged me from doing so. It is true that there are some cultural sensitivities that any foreigner must be aware of and have respect for. However, with the right attitude and motivation to get things done, anything is possible. I would encourage more people to look to growing their businesses in the MENA region. As far as the long-term strategy for Globelist goes, we want to spend the next five to seven years focusing on growing our business in MENA. After that, we will begin to reconsider the Southeast Asian markets, as they continue to grow and attract foreign investment, thus leading to lifestyle changes. There are many opportunities in emerging economies because they are vastly underserved. It is an exciting and dynamic place with a lot that still needs to be done.

It could not be a more exciting time to be a woman in technology, especially outside the West. The world is more connected than ever before, and there are so many people, both women and men, out there doing amazing things. Go ahead, take the risk, and start a company in a country you've never been to but have studied extensively. At worst, you will meet incredible people and have a great story, and at best, you will unlock an unparalleled opportunity with in-built barriers to entry that will help you retain a competitive advantage in your market.

References

[1] Diana Report, Women Entrepreneurs 2014: Bridging the Gap in Venture Capital, Professors Candida G. Brush, Patricia G. Greene, Lakshmi Balachandra, and Amy E. Davis, September 2014, page 7.

[2] "Female Founders on an Upward Trend, May 26, 2015" by Gene Teare, Ned Desmond.

[3] "Four factors That Predict Startup Success, and One That Doesn't", by Tucker J. Marion, Harvard Business Review, May 3, 2016.

[4] "Investors Prefer Entrepreneurial Ventures Pitched by Attractive Men," by Laura Huang, Harvard professor Alison Wood Brooks, and Sarah Wood Kearney and Prof. Fiona E. Murray, both of MIT's Sloan School of Management.

[5] "Female Founders on an Upward Trend", Crunchbase, by Gene Teare and Nes Desmond, May 26, 2015.

PART IV

Education for the 21st Century

Introduction

Women are notoriously underrepresented in STEM. A key way to increase their representation is to ensure that educational opportunities and professional support are properly invested in order to rectify this imbalance. This may seem to be an obvious statement, but it is the central tenet of this mission: to get more women into STEM positions in order to help create more entrepreneurial opportunities that supports everyone globally.

Exciting endeavors such as President Obama's Computer Science for All Initiative and FIRST, an international program designed to engage students K–12 in robotics and research, are giving us hope for a new generation of girls and young women engaged in STEM. However, we know that this is not enough. The real transformation will happen as a result of the follow-through, the perseverance, and the stamina of those involved after the programs fade from the news. We must approach this investment as vital for the survival of our species. This new generation is working on technologies such as keeping our identities secure and engineering infrastructure for clean water.

In this section, Shraddha Chaplot tells a love story of how she fell for numbers as a child and learned to use every part of herself (especially the artistic) to help create a culture of cool around STEAM for future generations. Sandra Bradly, the Director of the Internet of Things Lab at the University of Wisconsin-Madison, discovers collaboration and the power of "playing in the sandbox." Ankur Kamar calls for women everywhere to carry the torch for the next generation.

Across the world, women face significant difficulties in educating themselves. In Pakistan, Iffat Gill tells of the transformative power of her father's support for her wish to pursue computer science and ultimately face down archaic cultural limitations in order to help others achieve the same. Rwandan-born Alphonsine Imaniraguha triumphs despite living through the atrocities of genocide, PTSD, and an interruption in her secondary education.

All of these women have used considerable tenacity to get to where they are. While their stories inspire, they also demonstrate the obstacles women and girls face in getting an education, and the crucial role that mentors, parents, and community members can play in making that happen. Our call to action is clear: we must continue to light the path for the next generation's success. The change has already begun, and we cannot go back.

5

Women in Academia: A Potential STEM Powerhouse

By Nada Anid Ph.D.

Note: While this chapter focuses on the United States, much of it is germane to other countries, too.

5.1 What a Gender-Blind STEM World Could Look Like

Technology is shaping our daily lives, and inexorably, our futures. Yet in classrooms in the United States and around the world, STEM subjects are still widely perceived as academic disciplines to be "studied," not as tremendously powerful tools to be mastered for the purpose of solving humanity's problems and safeguarding life on earth. STEM professionals who wield these tools remain generally underappreciated.

Women remain underrepresented among these professionals. The percentage of women acquiring STEM degrees, holding faculty and administrative positions in STEM academic departments, and participating in the STEM workforce—once tiny—has been growing. But it still falls far short of what it can be.

What will our society look like when STEM and STEM talent are more highly prized, when many more girls and women aspire to be STEM professionals, when STEM fields become gender blind, and when more STEM talent is unleashed?

Here's a partial sketch

In popular culture—film, TV, videos, web sites, books, magazines—STEM experts are celebrated. Stories about women in STEM professions abound. Those women serve as role models for girls.

> *STEM teachers are the rock stars of K–12 classrooms, generating excitement and inspiring engagement. STEM subjects are perceived as powerful tools, and students are eager to master and use them.*
>
> *Science clubs rival football teams in fan appeal. (Well, not really, but their appeal is growing!)*
>
> *Young women in large numbers pursue STEM degrees. So do older women seeking to switch careers. Within academia, large cohorts of female students become their own change agents, dissolving barriers to learning and advancement. Many women become faculty and administrators.*
>
> *On the basis of sheer merit, female graduates enter STEM professions and advance within them. Many women hold leadership positions. Many women become patent holders. Many become entrepreneurs and job creators.*
>
> *In academia, industry, government, and non-profits, strategic plans to promote diversity become obsolete, thus liberating time, effort, and talent for other pursuits.*
>
> *The large influx of female STEM talent is an economic stimulus that strengthens our national security.*
>
> *Widespread basic STEM literacy and related critical thinking skills strengthen our democracy.*

This world may be within our reach, and the following chapters in this part explore the work of those making incredible strides globally. Later in the chapter, I offer a view point on this issue as an academic and provide recommendations.

5.2 What Is at Stake?

An Urgent National Need

The STEM workforce is rapidly becoming an ever-more vital national asset—essential to commerce, industry, academia, and government at all levels. The United States has an increasingly urgent need for a workforce equipped with STEM expertise—not only to remain economically competitive, but to solve pressing problems and reduce risks—some closely interrelated—that threaten health and safety, the environment, and national security.

Among them are climate change and other forms of environmental damage; the need for cheap, clean, reliable energy; crumbling infrastructure; cybersecurity vulnerabilities within businesses and government; and cyber-terrorism that could shut down our power plants and take down our grid.

Even as demand for STEM expertise is soaring, baby boomer retirements have been depleting the STEM workforce. And compared with China and India, the number of STEM professionals the United States is producing is tiny.

So urgent is labor-market demand that the shortage has become a public-policy issue that is being addressed at the highest levels of government.

An Unprecedented Opportunity for Women

As the STEM job market expands, it's offering women unprecedented opportunities to pursue rewarding careers that contribute to the greater good. Many initiatives are underway to capitalize on this, in the name of simple gender equality.

But along with equality, diversity is emerging as a compelling goal, for pragmatic reasons: evidence is rolling in that teams that are diverse in terms of gender, race, ethnicity, and other characteristics outperform those that aren't. (A recent Morgan Stanley study of the performance of 1,600 companies over 10 years found that those that have gender-balanced workforces and offer equal pay and work-life balance programs tend to deliver better returns on equity.)[1]

Not extending opportunities to women is inequitable. Not taking advantage of what they can contribute is an extravagance we simply can't afford.

A Precondition for a Healthy Democracy in a Complex, Risky World

Closing the STEM gender gap matters far beyond the STEM marketplace. Basic computer literacy is becoming as fundamental to many workplaces as the ability to read and write. And managers at all levels need a basic understanding of the technologies their organizations run on.

More profoundly, more-effective STEM education from early grades on can pay tremendous civic and societal dividends. In our increasingly complex, interdependent, technological society, risks of many kinds are rising to unprecedented levels. Citizens who possess basic scientific and digital literacy and strong critical-thinking skills are vital to a healthy democracy, sound decision-making, and our prospects for achieving a sustainable future.

A Grave Caution about Root Problems

While this chapter focuses on higher education, the STEM gap cannot be understood in isolation from K–12 classrooms. Despite many impressive initiatives to engage girls' interest in STEM studies, schools seem to be still failing on a massive scale, thereby depriving girls of the chance to experience the thrill of scientific discovery and innovation, and of career opportunities that could be richly rewarding. Here are a few key points to consider in this regard:

- *Cultural forces can suppress girls' interest in STEM.* Attitudes that girls internalize early on are too often reinforced in school. They can linger on in institutions of higher education. They can persist in the workplace.

  Girls' attitudes are shaped and reinforced by messages from popular culture and media—implicit but powerful—and by adults' expectations. What many girls learn early on is that *STEM is for boys.* Female STEM role models remain scarce in popular culture, which is largely still dominated by celebrities in sports, entertainment, music, and fashion. (The emerging STEM toy category is still marginal.)[2] TV shows and films featuring "nerds" and "geeks" are still largely male-dominated. The fragmentation of media further submerges STEM. The role of *women* in STEM is largely obscured.[3]

- *K–12 educational constraints are severe.* Some elementary teachers may unconsciously communicate their own discomfort with math. Low pay scales for teachers can't attract much top STEM talent. Strong labor-market demand forces other employers to pay high salaries. Ironically, this siphons off the potential teaching talent that might attract more girls to STEM, and thus help ease the labor shortage.

  Many STEM teachers diligently do their best but aren't equipped with all of the knowledge and skills they need. Bureaucratic *teach to the test* pressures divert capable teachers from genuine teaching. Formal requirements that teachers hold advanced degrees may bar from the classroom capable individuals who could ignite students' interest.

- *Cultural forces and classroom constraints converge.* In STEM classrooms, for a variety of reasons, girls tend to be less confident and much more risk-averse than boys, slower to speak up, and more afraid of failing. Thus they underperform, thus seeming to further confirm the assumption that STEM is for boys. The consequence is that, by high school, many girls have drastically narrowed the career options they imagine for themselves.

This convergence of root causes has had profound consequences for the lives of countless girls and women, and for society. What rewarding careers have

5.2 What Is at Stake? 127

been foregone? What discoveries and innovations could have been made if only...?

The classroom—day by day—is the crucible in which negative cultural influences can be overcome, minds engaged, and imaginations set racing. We cannot transcend the level of talent, skill, and knowledge that the classroom teacher is able to supply. Put bluntly: it may be magical thinking to assume that many more girls will pursue STEM studies if we don't *substantially* upgrade the caliber of K–12 STEM teaching.

This societal challenge is immense, and a new factor may be further complicating it: the extent to which the digital world is supplanting the physical. The Internet is a tremendous resource, but young digital natives who spend hours absorbed in screens are, to that extent, divorced from the physical world. Will this further weaken their engagement with STEM?

A Closer Look at That Hot Labor Market

> *Note*: STEM is a broad category, and definitions of the specific disciplines it spans vary. This chapter's narrow focus is engineering and computer science. But for brevity's sake, and when a wider context is referenced, the term "STEM" is often used.

For a quick sense of the scale of the "science and technology" workforce as at 2013 (the most recent data), glance at Highlights in the National Science Foundation Science and Engineering Indicators 2016.[4]

A narrower focus comes from the Bureau of Labor Statistics (BLS), which every two years projects labor-force levels for the next ten. Its most recent occupational-employment projections were published in December 2013. BLS categories don't neatly track ours, but they are still valuable indicators. BLS projects growth from 2012 through 2022 in these categories:

- Computer and mathematical occupations: 18%
- Architecture and engineering occupations: 7.3%

We also read the following on the BLS website:

> By 2022, [computer and math] occupations are expected to yield more than 1.3 million job openings...
>
> Software developers and programmers are expected to add 279,500 jobs by 2022, accounting for about 4 out of 10 new jobs in the... group... the rate of growth for information security analysts is expected to be 36.5 percent, making this the fastest-growing occupation in this group.[5]

We also read:

> The median annual wage ... in May 2012 was $76,270, more than twice the median annual wage for all wage and salary workers ... and the second highest of any major occupational group ... computer and information research scientists[6] and mathematicians, had median wages of more than $100,000 per year.[7]

There is a high demand for information-security analysts, who:

> plan and carry out security measures to protect an organization's computer networks and systems. Their responsibilities are continually expanding as the number of cyberattacks increases ...
>
> Most ... positions require a bachelor's degree in a computer-related field.
>
> The median annual wage ... was $90,120 in May 2015.
>
> Employment ... is projected to grow 18 percent from 2014 to 2024, much faster than the average for all occupations.[8]

As the BLS observes, cybersecurity, the implementation of electronic medical records, and the growing use of mobile technology will drive demand for software developers, programmers, and information-security analysts. Another driver: the ballooning quantities of data generated every second in an increasingly sensor-networked world.

The most recent salary survey from the National Association of Colleges and Employers reports that average computer-science salaries rose by 6.1% in 2014 to $62,100. The average engineering salaries rose 1.3% to $62,900.[9]

In its architecture and engineering category, BLS notes:

> Engineering occupations make up about two-thirds of this ... group, and they are projected to add 136,500 jobs over the next decade. Civil engineers will add 53,700 jobs by 2022 ... the most of any engineering occupation. Demand for infrastructure to provide services such as clean drinking water and waste treatment systems will drive job creation for civil engineers.[10]
>
> In May 2012, the four highest paying occupations in this group were all engineering jobs that typically require a bachelor's degree: petroleum engineers ($130,280)[11], nuclear engineers ($104,270)[12], aerospace engineers ($103,720)[13], and computer hardware engineers ($100,920).[14]

Clearly, demand is strong, and opportunities are rich.

5.3 The Gateway to Opportunity: Academia in Flux

Characterizing the Field

The gateway to that hot STEM labor market is, still, largely academia.

Girls who do choose to pursue STEM studies enter a world of learning and credentialing, in which a formal curriculum will lead them to degrees that qualify them for professional positions, and more broadly equip them to pursue and create careers. For some graduates with advanced degrees, academia becomes an employer, hiring them for teaching or administrative positions.

For many years, academia was an orderly world in which disciplines were neatly siloed, and today some of its features remain stable, even static. But other features are now in play. Notably, over the past few decades, STEM fields have become far more dynamic, with researchers pursuing more cross-disciplinary collaborations and forging hybrid disciplines.

There are two primary sources of data:

- The American Society for Engineering Education (ASEE) annually publishes extensive data on "engineering and engineering technology" colleges. ("Engineering technology," which may seem redundant, emphasizes hands-on applications rather than mathematical and design fundamentals.) ASEE's most recent profiles cover 358 colleges and universities that offered engineering programs for the academic year 2014.[15]
- Its Engineering by the Numbers report tracks enrollment and degrees awarded for subdisciplines, tenured and tenure-track faculty, research expenditures, and more. To indicate scale, here are a few data points for 2014 (with gender breakdowns reserved for later on in this chapter):

   Total undergraduate enrollment*: 569,274
   Bachelor's degrees*: 99,173
   Master's degrees awarded*: 51,690
   Doctoral degrees awarded*: 11,309
   Total number of "teaching personnel": 31,777

   *Does not include computer science taught outside engineering schools and departments.[16]

- The Computer Research Association[17] produces a vast trove of data.[18] Its annual Taulbee Survey documents trends in student enrollment, degree production, employment of graduates, and faculty salaries in academic entities that grant PhDs in computer science, computer engineering, or information in the United States and Canada. (The inclusion of Canada complicates the use of some data.) Data on bachelor's and master's degrees are also reported.[19]

Academic institutions, both public and private, are extremely heterogeneous and range in size from small to huge. (MIT's operating budget for fiscal year 2015 was over $3 billion.) Public institutions receive taxpayer funding; all receive revenues from tuition, contracts with government agencies and businesses, and the licensing of intellectual property. Funding also flows from foundation grants and philanthropic gifts.[20]

Academic departments operate within bureaucracies but can also be entrepreneurial, routinely partnering and collaborating with industry and government, typically conducting research that their partners fund. Borders can be porous, with some faculty both teaching and working as consultants to, or employees of, industry and government. Within institutions, faculty may be tenured or adjunct. The current trend is toward hiring a substantial proportion of adjunct faculty, for reasons of cost and flexibility.

The advent of massive open online courses (MOOCs) is a potential game-changer in some fields. Moreover, given strong labor-market demand for computer science, what's learned from bootcamps and crash courses such as Codecademy's, which are online and free, and even the raw talent and skills of self-taught coders, may be enough to gain employment, or even launch a company, without the benefit of a formal degree. These developments may have long-term consequences for academia.[21]

Beyond individual institutions lies the busy world of professional associations. Their many events offer opportunities for networking, and with their publications, help advance STEM fields and careers.

Title IX Protections Against Gender Discrimination

Because the central concern of this book is girls' and women's status and opportunities, the importance in academia of legal prohibitions against gender discrimination must be mentioned. In the United States, gender discrimination is illegal under federal, state, and local laws and regulations, including Title IX of the Federal Education Amendments Act of 1972. The United States Department of Education says the following in this regard:

> Title IX . . . prohibits discrimination based on sex in education programs and activities that receive federal financial assistance.
>
> Examples include . . . discrimination in a school's science, technology, engineering, and math (STEM) courses and programs.[22]

That prohibition is comprehensive, thus protecting students, employees, and people applying for admission and employment.

If any part of an educational entity receives any federal funds for any purpose, all of its operations are covered by Title IX, and the entity must designate at least one employee to coordinate efforts to comply with it. That is, every entity that receives federal funds for any purpose must have a Title IX coordinator, and that official's duties extend to STEM.

5.4 What Are Women up Against?

Cautions about Data

To venture onto the Internet in search of *current* data about the status of women in STEM is to discover that determining status and trend lines is not altogether straightforward. While data are proliferating, some sources present problems.

One problem concerns the recency of data. Some current publications, and news stories about them, cite old statistics and research data, but not for historical purposes. Rather, in academia, the pace of publication can be slow. Similarly, large-scale reports published by NGOs and government agencies can take several years to assemble. Some online stories bear no date at all. Thus, close attention to dates—to when research and survey data were collected, and to dates of publication—is advisable.

Another problem arises from variations in how STEM is defined. "Technology" is a hopelessly broad term. "STEM" is narrower, but the subdisciplines it spans can vary by source. For example, some sources include life sciences and social sciences. The American Society for Engineering Education tracks a whopping 23 engineering subdisciplines that include chemical, mechanical, civil, computer, electrical, and biomedical engineering. And, as noted above, the dynamism of research in some fields is also spawning new subdisciplines. For example, data analytics is an emerging field within computer science that also figures in business, marketing, finance, and even health science. The National Science Foundation Science and Engineering Indicators 2016 report discusses the 2013 "science and technology" workforce, defining it broadly.[23]

Consequently, depending on how a field is defined, different sources may report somewhat different statistics for the same year on the percentage of women who are pursuing or earning degrees and holding jobs.

Nonetheless, a sampling of data offers valuable insights into how girls and women are faring in STEM fields. For context, we'll start with a look at the larger workplace culture, and then sample some indicators of women's status in academia, as well as a number of marketplace indicators.

132   Women in Academia: A Potential STEM Powerhouse

Data sources that span more than one sector can't be neatly categorized, so the organization below is somewhat rough, and because of the diverse nature of sources, the effect is scattershot. Nonetheless . . .

Within the Larger Workplace Culture

Let's start at the top: A 2016 report by the American Association of University Women, *Barriers and Bias: The Status of Women in Leadership*, "examines the causes of women's underrepresentation in leadership roles in business, politics, and education and suggests what we can do to change the status quo."[24] The one-page summary indicates that, while "blatant sex discrimination is still a problem,"

> . . . subtler problems like hostile work environments, negative stereotypes about women in leadership, and bias still keep women out. Unconscious or implicit bias can cloud judgment.[25]

For women aiming to advance toward and break through the glass ceiling, a comprehensive 2015 study from McKinsey and Lean In, Women in the Workplace, finds that "women face greater barriers to advancement and a steeper path to senior leadership." Its stunning "corporate-talent pipeline" graphic depicts a decline in the percentage of professional employees who are women from 45% at entry level to 17% at "C suite" level.[26]

Harvard University researcher Iris Bohnet, author of a 2016 book, *What Works: Gender Equality by Design*[27], includes this sobering reflection in a March 11, 2016, *Wall Street Journal* piece, "Real Fixes for Workplace Bias":

> It is hard to ignore the possibility that all the time and money devoted in recent decades to promoting diversity at our major institutions has largely been wasted.[28]

Majora Carter[29], an urban activist based in New York City and who is a graduate of the Bronx High School of Science and a member of the Advisory Board of the Bronx Academy of Software Engineering High School, put it memorably: "There's no such thing as a glass ceiling—it's actually just a thick layer of men."[30]

A mesmerizing, lava-lamp-like animated graphic produced by the World Economic Forum displays how gender differences in the United States workplace have evolved in terms of employment and wages. "Math/Science" is one category.[31]

Tantalizingly, McKinsey's Women Matter[32] web page features an April 16, 2016, report, The Power of Parity: Advancing Women's Equality in the United States, and says,

Every state and city in the United States has the opportunity to further gender parity, which could add $4.3 trillion to the country's economy in 2025.[33]

STEM Patterns and Trends in Academia

Change the Equation "works at the intersection of business and education to ensure that all students are STEM literate by collaborating with schools, communities, and states."[34] Its "Stemistics" present many data bytes on a variety of topics. For example, we read:

> Although women receive about 36% of STEM Ph.D.'s, they make up about 18% of full professors in science and engineering.[35]

A January 10, 2014, news story[36] reported on an analysis of 2013 data that found that in Mississippi and Montana, no female African American or Hispanic students took the Advanced Placement exam in computer science.

In fact, no African American students took the exam in a total of 11 states, and no Hispanic students took it in eight states.[37]

> ... about 30,000 students total took the AP exam for computer science ... Less than 20% ... were female, about 3% were African American, and 8% were Hispanic.[38]

That story presents a dramatic graphic of the number of female test-takers by state. Will more-recent data reveal much change?

ASEE's March 2016 *Connections*[39] newsletter presents this table comparing data for selected BA degree programs between 2005 and 2014:

Disciplines with the Highest and Lowest Female Enrollment				
Discipline	2005	2008	2011	2014
Environmental Engineering	38.3%	41.7%	44.0%	47.4%
Biological Engineering and Agricultural Engineering/Biomedical Engineering	36.8%	36.9%	38.7%	41.9%
Chemical Engineering	34.2%	33.3%	31.7%	33.5%
Mining Engineering	16.9%	12.7%	14.7%	14.9%
Computer Science (Inside Engineering)/Computer Engineering/Electrical Engineering	11.0%	10.8%	11.5%	13.4%
Mechanical Engineering	10.8%	10.8%	11.9%	13.8%

©American Society for Engineering Education, *Connections*, March, 2016; *Profiles of Engineering & Engineering Technology Colleges, 2014.*

The extremes are striking: women constituted 47.4% of students enrolled in Environmental Engineering and 13.4% of those enrolled in Computer Science (inside Engineering)/Computer Engineering/Electrical Engineering.

Here are other ASEE data for 2014 (reformatted for our purposes):

Female Engineering Students

Women received a higher percentage of **engineering bachelor's degrees** again, for the sixth straight year, climbing from 17.8% in 2009 to 19.9% in 2014. Based on enrollment trends, we expect to see the percentage of women receiving an engineering bachelor's degree to continue to increase slightly over the next few years.

The percentage of **engineering master's degrees** that went to women also rose in 2013, with 24.2% of all such degrees going to women. While an all-time high, this represents a mere two-and-a-half-percentage-point increase over 2005, and continues a stable trend of the last 10 years.

Women earned 22.2% of **doctoral degrees**, a slight decrease from 2013, but in the decade since 2005 they have seen their share grow by almost 5%. We can expect the percentage of doctoral degrees awarded to women to remain about the same over the next few years.

Faculty

Women comprised 15.2% of **tenured and tenure-track faculty** as of fall 2013, thus continuing a slow increase since 2001, when they represented 8.9%.

The percentage of **full professors** who are women increased from 9.4 in 2013 to 10.2 in 2014.

The percentage of **associate professors** who are women increased from 17 in 2013 to 18 in 2014, while the percentage of assistant professors who are women remained the same at 22.8 over the same period.

The National Student Clearinghouse Research Center's January 2015 Snapshot Report, Science & Engineering Degree Attainment: 2004–2014, presents data on degrees, based on National Science Foundation (NSF) classifications, which include social sciences and psychology in "science." Here are some data that exclude those disciplines.[40]

Girls Who Code offers interesting data (but doesn't cite sources or dates):

In 1984, 37% of all computer science graduates were women, but today that number is just 18%. Twenty percent of AP Computer Science test-takers are female, and 0.4% of high school girls express interest in majoring in Computer Science. What's going on?

		Women's Share in 2004	Women's Share in 2014
Engineering	Bachelor's	20%	19%
	Master's	21%	24%
	Doctoral	19%	23%
Computer Science	Bachelor's	23%	18%
	Master's	30%	27%
	Doctoral	21%	21%

While 57% of bachelor's degrees are earned by women, just 12% of computer science degrees are awarded to women.

In middle school, 74% of girls express interest in . . . [STEM], but when choosing a college major, just 0.4% of high school girls select computer science.

Despite the fact that 55% of overall AP test takers are girls, only 17% of AP Computer Science test takers are high school girls.

Women make up half of the United States workforce, but hold just 25% of the jobs in technical or computing fields.[41]

In a lively February 24, 2016, blog post, CloudCalc, Tom Van Laan writes,

> The percentage of women among engineering students in the United States has climbed over the past 25 years, to a "high" of approximately 19%. The bad news? Over 40% of those women engineering graduates either never enter the engineering profession, or drop out of that profession within a few years (a rate four times as high as that of men), reducing the percentage of women in the engineering work force to a paltry 11%. The software development world nearly mirrors those stats—18% of computer science students are women, but recent data has shown that women hold less than 15% of the technical positions in Silicon Valley.[42]

Alarming statistics come from a new initiative, Women in Technology and Entrepreneurship in NY (WiTNY):

- By 2018, the United States will be graduating only 52% of the needed technology workforce from our universities. In New York, the tech job market is growing four times faster than any other industry job market.
- While the number of women attending college is at an all-time high, less than 1% of women graduate with degrees in technology-related disciplines, compared with 6% of men.

- Over the past 20 years, the percentage of computer-science degrees awarded to women has been declining steeply—from 37% to 18%.

In 2016, there are 350 engineering-school deans in United States universities, and a mere 10% of these are women.

Here's another trend that presents a challenge: the growing number of STEM students—of both genders—who speak, read, and write English as a second language. Language barriers can hamper learning and handicap writing skills, and thus erode academic-retention rates.[43]

STEM Patterns and Trends in the Marketplace

Employment

The National Center for Women & Information Technology reports that in 2015, "25% of the computing workforce were women, and less than 10% were women of color (5% were Asian, 3% were African American, and 1% were Hispanic)."[44]

National Science Foundation Science and Engineering Indicators 2016 reports as follows:

> Women remain underrepresented in the S&E workforce, but less so than in the past . . . women constituted 50 percent of the college-educated workforce, 39 percent of employed individuals whose highest degree was in an S&E field, and 29 percent of those in S&E occupations.

It adds that women held 62% of jobs in social sciences and 48% in life sciences, but only 15% in engineering, 31% in physical sciences, and 25% in computer and mathematical sciences.[45]

The American Association of University Women's report Why So Few? Women in Science, Technology, Engineering, and Mathematics, published in 2010, offers advice that still seems valid:

> To diversify the STEM fields we must take a hard look at the stereotypes and biases that still pervade our culture. Encouraging more girls and women to enter these vital fields will require careful attention to the environment in our classrooms and workplaces and throughout our culture.[46]

A February 2016 MIT *Technology Review* story, Female Coders Are "More Competent" Than Males, According to a New Study reports on a study of 1.4 million computer programmers that "suggests women are better programmers but face persistent gender bias."[47]

Entrepreneurs

A study published by the National Bureau of Economic Research in 2012 reported that 5.5% of commercialized patents are held by women.[48]

Male entrepreneurs still far outnumber women. An April 2016 *MIT Technology Review* story on the challenges women face reports the following:

> The biggest issue female founders are encountering is fund-raising, as most venture capital investors continue to be men...[49]

That story points to an April 2016 *Fortune* magazine article, Venture Capital Still Has a Big Problem With Women, that says, "Fewer than 6% of all decision-makers at United States venture capital firms are women."[50]

An August 4, 2015, White House news release on its "Demo Day" promotion of inclusive entrepreneurship reports the following:

> Just three percent of America's venture capital-backed startups are led by women, and only around one percent are led by African-Americans . . . about four percent of venture-capital investors based in the United States are women.

But an August 2014 article in *The Atlantic*, drawn from a study published in 2014, presents striking findings about women's success with Kickstarter. In funding tech projects, women had twice the success rate of men. The researchers conclude that women who want to promote women in male-dominated industries are likelier to support projects headed by women.[51]

Initiatives Underway

The data cited above point straight to obvious measures:

- *Female students*: Recruit them, retain them, and help them find jobs in STEM fields. Beyond "jobs," help them aspire to careers, and entrepreneurship.
- *Female faculty and administrators*: Recruit them, retain them; ensure equal pay and opportunities for advancement.

Taken together, root problems and challenges are not to be underestimated. But are the forces of history with us? As labor-market demand for STEM talent soars, concern about women's continued marginal status in that work-force grows. A multitude of parties—in academia, public and private sectors, and the media—are accelerating and expanding efforts to close the STEM gap. Their dynamism is exhilarating—in fact, historic.

Crucially, efforts that benefit women can benefit minorities, and vice-versa. And of course many women are minorities. Efforts to recruit minorities to STEM studies and help them advance in the workplace date back decades, and are gaining force.

Gender-gap initiatives range from narrowly focused to broad-spectrum. Here, too, neat categorizations aren't always possible. Roughly grouped, below is a sampling.

Addressing Root Causes
Not Gender-Specific, but Benefiting Girls

- President Obama's Computer Science for All Initiative, announced in his 2016 State of the Union address, aims to equip K–12 students with "the skills they need to be creators in the digital economy . . . and to be active citizens in our technology-driven world."[52]
- NYC's Computer Science for All aims to ensure that all of New York City's 1.1 million public school students "have access to a high-quality computer-science education that puts them on pathway to college and career success." The 10-year, $80-million plan is a partnership between the city and the private sector. San Francisco's Board of Education voted to offer computer science from pre-K through high school and to make it mandatory through eighth grade by the 2016–17 school year. Chicago intends to make a year-long computer science course a high-school graduation requirement by 2018.[53]
- In 2013, the NSF announced a United States-Finnish collaboration[54] to improve STEM education in K–16 classrooms, through eight research projects conducted through Science across Virtual Institutes (SAVI).[55]
- FIRST works internationally to create *"a world where science and technology are celebrated and where young people dream of becoming science and technology leaders."* It engages K–12 students in research and robotics programs, and expected to reach more than 400,000 students in the 2014–2015 school year. Its 2014 Annual Report states that "almost 90 percent of FIRST Alumni are now studying for or working in STEM careers, and female FIRST participants are four times more likely to study STEM subjects due to participation."[56]

  FIRST's founder, Dean Kamen, and NASA's administrator spoke about STEM education at NYIT's commencement and at the Brookings Institute in May 2016.[57] View the videos here.[58]

5.4 What Are Women up Against? 139

- Published in 2008, a National Academy of Engineering report remains—sadly—timely and valuable: Changing the Conversation: Messages for Improving Public Understanding of Engineering.[59]

Directly Engaging Girls
- The National Center for Women & Information Technology comprises more than 650 universities, companies, non-profits, and government organizations that aim to increase women's participation in computing and technology. Its scope includes K–12, higher education, industry, and entrepreneurial careers.[60]
- MIT works aggressively to engage girls in K–12 in STEM. Examples:
    o The MIT Women's Technology Program is a rigorous 4-week summer academic and residential program for female high-school students who explore engineering through hands-on classes, labs, and team-based projects in Electrical Engineering and Computer Science, or in Mechanical Engineering.[61]
    o The MIT K–12 Outreach Initiative advances STEM literacy through programs for elementary, middle, and high school, and engagement with after-school programs, students, and families. It makes STEM education accessible to underperforming and underrepresented groups, including girls, and offers resources for teachers.[62]
- On a smaller scale, and with equal determination, my own institution, the New York Institute of Technology, works to close the gender gap. This brief video interview explains our approach. Our efforts include the following:[63]
    o Introduce a Girl to Engineering Day celebrates National Engineering Week. We host girls aged 12–18 to increase their awareness of engineering and pique their interest in careers in engineering and computer science.[64]
    o Our NYIT Engineering and Technology Showcase Competition invites teams of high-school juniors to exhibit projects based on science and/or engineering principles. Each team must include at least one female student. Cash and certificates are awarded.[65]
- Carnegie Mellon's School of Computer Science offers extensive STEM outreach programs that "pay particular attention to encouraging more participation by women and under-represented minorities." Programs

- reach middle and high-school students, and train high-school teachers and college instructors in how to teach computer science.[66]
- The American Association of University Women pursues a vigorous, multi-pronged approach to promoting STEM education for girls and women.[67]
- *Huffington Post*'s Girls in STEM Mentorship Program offers many web pages loaded with content that's fun to explore.[68]
- Girls Who Code aims "to inspire, educate, and equip girls with the computing skills to pursue 21st century opportunities." It exposes girls "to real life and on screen role models. We engage engineers, developers, executives, and entrepreneurs to teach and motivate the next generation."[69]
- Most encouragingly, young women themselves are leaning into this effort. For example:
    o Stanford University students launched she++ to overcome the stereotype that computer science is not for women and to advance underrepresented groups in the technology field. It works with high-school and college students, and works with industry to create a culture that is more appealing to women.[70]
    o The Twitter campaign #ILookLikeAnEngineer is a breath of fresh air![71]

Powering up Academia and Professions

- As Matthew Flamm reported in Crain's New York Business on March 21, 2016:[71]

    In partnership with Cornell Tech and Verizon, and with help from other tech companies, the City University of New York will launch an introductory computer-science class designed to appeal to women and increase that population at the start of the pipeline. The class is part of a wider initiative called Women in Technology and Entrepreneurship in NY, or WiTNY, which will include internship and mentoring programs in coordination with local companies and tech leaders.

    The introductory class . . . teaches computer science not as an abstract discipline but as a subject that can apply to broader issues and concerns.

    . . . "Imagine teaching robotics not for the sake of robotics but to help with disabled-access issues. Imagine teaching data analytics to track the path of refugees moving from Syria to Europe."

5.4 What Are Women up Against? 141

- The White House Women[73] in Stem page links to information sources, and tells us the following:

    The Office of Science and Technology Policy, in collaboration with the White House Council on Women and Girls, is dedicated to increasing the participation of women and girls—as well as other underrepresented groups— . . . by increasing the engagement of girls with STEM subjects . . . encouraging mentoring to support women throughout their academic and professional experiences, and supporting efforts to retain women in the STEM workforce.[74]

- The American Society for Engineering Education's broad portfolio includes the following statement:

    The Engineering Deans Council . . . works to promote diversity and inclusiveness in all aspects of engineering education, research, and engagement.[75] ASEE declared academic year 2014–2015 the Year of Action in Diversity. In March of 2016, deans of engineering schools . . . signed a letter committing their institutions to steps to advance women and other underrepresented groups in STEM fields.[76]

- The Society of Women Engineers, formed in 1950, was the first society dedicated to advancing women engineers. With nearly 30,000 members today, it offers many activities, including a diversity and inclusion program and a K–12 program.[77]
- IEEE Women in Engineering is the largest international professional organization dedicated to promoting women engineers and scientists and to inspiring girls around the world to pursue careers in engineering. "Change the world—become an engineer" is its exhortation to girls.[78]
- Computing Research Association's committee on women provides "mentoring and support for women at every level of the research pipeline: undergraduate students, graduate students, faculty, and industry and government researchers." Its mission extends to engineering as well as computer science.[79]
- Women Who Code[80] helps women excel in technology careers. Initiatives include free technical training in programming languages, the facilitation of networking and community building, and advocacy to raise the profile of women in tech fields. It advises companies on hiring practices for an inclusive workplace.

142 Women in Academia: A Potential STEM Powerhouse

- Code.org works to increase diversity in computer science and to establish it as part of the core curriculum. It works "across the education spectrum: designing our own courses or partnering with others, training teachers, partnering with large school districts, helping change government policies, expanding internationally via partnerships, and marketing to break stereotypes."[81]
- The Anita Borg Institute is named for a computer scientist who in 1987 started a digital community for women in computing. Today, ABI works with women technologists in over 50 countries, and partners with leading academic institutions and Fortune 500 companies. ABI evolved from a belief Borg expressed as follows:

  > Around the world, women are not full partners in driving the creation of new technology that will define their lives. This is not good for women and not good for the world . . . Women need to assume their rightful place at the table creating the technology of the future.[82]

  Still true.

5.5 Strategies for Closing Academia's STEM Gender Gap

The strategies offered below—drawn from many sources, including initiatives described above and my own experience and my colleagues'—aren't definitive. But together they signal that much can be done, today, to narrow the STEM gender gap.

Because the difficulty of closing the gap varies from institution to institution, and changes over time, the strategies must be adapted to local conditions. Every member of an institution—administrators, faculty, students—can employ some of them.

Overarching Strategies
Learn, and Contribute
- Apply strategies proven to work—and sustain them, as Norman L. Fortenberry, Sc.D., Executive Director of the American Society for Engineering Education, stresses.
- To foster a gender-neutral culture, benchmark what the competition is doing. Figure out how to do it better.
- Inventory your institution's strengths and build on them.
- Toward the goal of continuous learning, assess failures and successes.

- Participate in and help advance some of the initiatives described above.
- Share best practices; practice them.

Strategize, Partner, and Collaborate

Opportunities for forging creative partnerships and collaborations in STEM fields—among academic, private sector, and public sector parties—abound. Pursue them; create them.

Beam STEM Messaging from the Top

- From the "board room," signal a loud, clear, consistent commitment to advancing women in STEM fields. Make this a strategic goal, in the name of equity and diversity, which boosts performance.
- Showcase that commitment online, in print, in person.
- Convey a sense of optimism and momentum.
- Consistently aim at a female target audience that is itself diverse.
- Use women as the face and voice of the institution as often as possible, as speakers at events, as panelists, as spokespersons.
- Exploit social media to get your message out.
- Stress the mission of "engineering for society."

Recruiting Female Students

Learn

Learn from institutions that have worked aggressively and successfully to increase female STEM enrollment,[83] such as,

- Harvey Mudd College[84], which "seeks to educate engineers, scientists, and mathematicians well versed in all of these areas and in the humanities and the social sciences so that they may assume leadership in their fields with a clear understanding of the impact of their work on society."
- The University of Washington, whose Computer Science & Engineering program "is widely recognized for . . . success in attracting women to the field."[85]
- Explore research on pragmatic tools for reaching K–12 girls, such as the NSF-funded project reported on in the paper Using an Extension Services Model to Increase Gender Equity in Engineering.[86] That project aimed to increase the number of women attaining baccalaureate degrees in mechanical and electrical engineering.

Fine-Tune Messaging

- Stress the extraordinarily wide applicability of STEM degrees, and labor market demand for STEM expertise. A video on WiTNY's web site makes this point powerfully.[87]
- Stress opportunities for travel and for working with many different kinds of people. Stress the dynamism of STEM fields and the promise that "you will not get bored."
- Contextualize the curriculum: show how pervasively STEM disciplines have shaped the world, how heavily dependent we are on STEM, why and how mastery of a discipline matters, how it relates to daily life.
- Stress the chance to make a difference.
- Shift the emphasis from mastering formal disciplines to STEM's real-world utility as a powerful tool, for solving, and preventing, serious problems:
    - Not "You can become an environmental engineer," but "Help develop innovative, low-cost, ways of removing pollutants like lead from drinking water, and see your technology spread around the world!"
    - Not "Learn how to write computer code," but "Help advance the frontiers of cybersecurity!"
- Convey the concept of scalability.
- Offer many real-world role models: women who have worked, and are now working, successfully in STEM fields. Start with the NSF's Pioneering Women in STEM.[88]
- Share stories of women pioneers who played pivotal roles in early software development, to correct the widespread misimpression that the field was entirely created by men. Start with legendary Grace Hopper, whose career included serving as a US Navy Rear Admiral and later leading teams that developed the first computer language compiler and the first user-friendly, business computer software program, COBOL.[89]
- Explain potential career paths: how professionals actually use STEM skills.
- Help girls and women imagine becoming entrepreneurs themselves.
- Stress a culture of collaboration and teamwork.
- Draw from this powerful case in framing your appeals to girls:
    - 5 Reasons Why Girls (and Boys) Should Consider a Career in IT[90]

- Design recruitment campaigns not only for young women but for older women who may want to change careers, or enter the workforce for the first time.

Stress Specific Benefits

- Offer scholarships. Successful alumni, both female and male, are potential benefactors. Beyond the gift of funding, women alums are role models, and some may become mentors.
- Promote the benefit of mentorships and internships. (More on these below.)
- Offer summer STEM camp experiences. (The Intrepid Museum GOALS for Girls summer intensive camp offers a model.)[91] Solicit corporate and foundation funding to underwrite scholarships for campers.

Consider Redesigning Courses and Curricula

- Design courses that help students more easily master foundational subjects they may be intimidated by.
- Learn from work being done by Stevens Institute of Technology under a 2015 NSF grant to transform STEM courses.[92]
- Consider recommendations in Achieving Systemic Change: A Sourcebook for Advancing and Funding STEM Education.[93] This 2014 report resulted from a workshop organized by the Coalition for Reform of Undergraduate STEM Education on how to transform undergraduate STEM education.
- As suggested in this blog post, design courses focused on solving real-world problems that may particularly appeal to women.[94]
- But should computer courses be designed differently for women? A July 2014 news story presents opposing views.[95] Two female computer science professors at Carnegie Mellon redid earlier gender studies of the appeal of real-world applications of technology. They found no gender differences. By contrast, Harvey Mudd College reframed engineering and computer science as creative problem solving, not hard-core programming.[96] For example, an unpopular course titled "Introduction to Programming in Java" became the very popular "Creative approaches to problem solving in science and engineering using Python."
- In considering course design, consider the context within which they're offered; for example, the availability of mentoring.

Retaining Female Students

- Carry forward strategies offered above.
- Assign faculty advisors to help students design a master course of study, and to modify it to serve their evolving needs and interests as they explore STEM fields.
- Arrange for mentors drawn from faculty, alumni, older students, and industry partners.
- Explore MentorNet, which "combines the technology of social networks with the social science of mentoring" to help mentors & protégés in 70 STEM fields connect across generational, gender, racial, cultural, and socioeconomic boundaries."[97]
- Engage alumni in arranging internships and—even better—workplace collaborations. NYIT's Entrepreneurship and Technology Innovation Center offers invaluable "before the job training."[98] Crucial to entrepreneurship training are learning strategies for overcoming the "fear factor," which makes girls and women risk averse. Also crucial: gaining the communication skills they'll need to make the case for their business ventures, and themselves.
- Foster a culture in which failures are understood as a normal part of learning and discovery. Help students shift their perceptions of failure, so that instead of fearing it, they exploit it as an opportunity to learn.
- In every course, connect the dots, stressing how a discipline matters in the real world.
- Structure courses around problem solving to build a sense of personal efficacy.
- Build a sense of community. Foster collaborative projects. Encourage women to help each other.
- Include a woman on every team of every kind whenever possible.
- To address English-as-a-second language problems, offer coaching to students who need help with writing and presentation skills.

Recruiting and Retaining Female Faculty

- Apply proven workplace-bias remedies. As Harvard University researcher Iris Bohnet writes in "Real Fixes for Workplace Bias,"

   > In recruitment: Be vigilant about the language you use . . . use gender-neutral wording. When hiring, learn from the blind auditions of symphony orchestras . . . Take advantage of new tools that easily allow firms to anonymize applicant information . . . Rely

5.5 Strategies for Closing Academia's STEM Gender Gap

on data to understand what is broken, and measure whether there are biases in how employees are supported in performing their jobs and evaluated for results.[99]

- Explore new online services, such as Textio, that can spot language problems.[100]
- Recognize the subtle, pervasive effects of implicit gender bias. Provide implicit bias training to all members of academic search committees.
    - But wait! Good news? "National hiring experiments reveal 2:1 faculty preference for women on STEM tenure track," published in April 2015 by the National Academies of Science, reports on a large-scale study conducted by University psychology researchers of gender bias in hiring.[101] Findings, telegraphed in the paper's title, suggest that, "it is a propitious time for women launching careers in academic science. Messages to the contrary may discourage women from applying for STEM . . . tenure-track assistant professorships." And, the researchers write,

        . . . the mechanism resulting in women's underrepresentation today may lie more on the supply side, in women's decisions not to apply, than on the demand side, in anti-female bias in hiring. The perception that STEM fields continue to be inhospitable male bastions can become self-reinforcing by discouraging female applicants, thus contributing to continued underrepresentation, which in turn may obscure underlying attitudinal changes.

- Offer paid family leave as a benefit.
- Include a woman on every team of every kind whenever possible.
- Learn what grantees have achieved under the NSF's program ADVANCE: Increasing the Participation and Advancement of Women in Academic Science and Engineering Careers (ADVANCE).[102]
- Adopt a diversity strategic plan, as recommended by the Diversity Committee of the US Deans Council of the ASEE. The work of developing such a plan is itself invaluable, for assessing conditions, articulating a vision, defining priorities, and assigning responsibilities and accountability for implementation.
- Data analytics is best suited to large organizations with large databases. Academic engineering schools and departments tend to be small in scale. But rigorously analyzing data whatever the scale remains essential. For example,

    o Audit pay by gender over a rolling 5-year period.
    o Track patterns of promotions over a rolling 10-year period.

Special Focus: The Crucial Role of Men

Because men dominate most STEM fields, they're well positioned to help girls and women advance, by advising, mentoring, sponsoring, and otherwise encouraging them. And many men recognize the merit of closing the STEM gender gap, and are willing to help. They should be encouraged to participate in initiatives.

Girls and women should be encouraged to identify men whose counsel can benefit them, and to ask for their help.

5.6 Academia's Own Learning Curve

Confronting Realities

Efforts to attract many more girls and women to STEM studies and professions date back decades, as do efforts to boost women's share of academic faculty and administrative positions. Many committed women, and men, have worked hard to advance these goals.

Why has progress been so slow? How much longer will it take to close the gender gap? The grave caution about root causes must be repeated:

- Implicit signals from popular culture coupled with adult expectations shape girls' belief that STEM is not for girls.
- In STEM classrooms, under-confident girls may underperform, reinforcing their belief.
- Many STEM teachers aren't equipped to ignite girls' interest in STEM subjects.
- School systems can't match salaries the private sector pays for STEM talent.
- By high school, many girls have drastically narrowed career options they imagine for themselves.

Moreover, ominous fiscal uncertainties cloud the future of US federal, state, and local governments and their ability to invest more heavily in education.[103]

But as our sampling of initiatives above indicates, greater commitment than ever before, from more sources than ever before—including the highest levels of government—is being devoted to closing the STEM gap and, more narrowly, the STEM gender gap.

Meanwhile, market forces, while diverting STEM talent from classrooms, are adding momentum. And as more students become code-literate, that literacy and related habits of mind may equip and motivate them to pursue more STEM subjects. The advent of the "maker" and "DIY" culture is a welcome stimulus, too. Progress does seem possible.

Asking Tough Questions

But accelerating progress requires asking tough questions. Some point directly to action steps; some to further research:

- Pay scales in public schools aren't likely to rise quickly, if at all: funding for public education may be imperiled as cities and states divert funds to meet pension obligations, repair aged infrastructure, and meet other pressing needs. What creative educational strategies can help compensate for public schools' inability to compete with growing private sector demand for STEM talent?
- How much would it actually cost to upgrade primary and secondary education nationwide enough to meet modest close-the-STEM-gap goals? Who could develop ballpark estimates?
- Finland is widely viewed as having the best education system in the world. What's being learned from the NSF sponsored US-Finnish SAVI collaboration?[104] From Finland's 2014 national program to strengthen STEM skills in students and help teachers teach?[105]
- What's being learned from Massachusetts, where school students outperform their peers nationwide in math and science?[106] And how much of its success is replicable in less affluent states?
- For the fields of engineering and computer science, we need one handy, comprehensive, accurate, up-to-date data baseline for tracking progress in closing the gender gap in K–12, academia, and the marketplace. Who should identify, evaluate, and assemble appropriate data sources? How should this project be funded? managed? promoted?
- Should faculty diversity be a criterion for school accreditation? For rankings of colleges and universities published by US News and World Report and other parties? If yes, how can this be achieved?
- Faculty are experts in their subject matter but typically have no formal training in how to teach effectively when they begin their teaching careers. "You may be teaching, but are they learning?" is a cautionary adage that ASEE's Executive Director, Norman L. Fortenberry, Sc.D., reminds us of.

Emily L. Allen, Ph.D., Dean of the College of Engineering, Computer Science and Technology at California State University, Los Angeles, observes that, "Resources for extensive training for new faculty may be an issue for many institutions. Requiring professional development for more experienced faculty may be even more challenging, especially in bargaining unit environments. Hiring institutions should consider selecting candidates who have had training during their PhD programs."

And, not all students know how to learn well.

A number of campus-based centers help engineering faculty and students better succeed.[107] How well are they helping women students? How can their STEM success stories be replicated widely and rapidly?

- For undergraduate and graduate STEM students for whom English is a second language, how severe a problem are language barriers? How adequately are institutions tackling this problem? What works best?
- In 2015, 73 percent of men and 80 percent of women were using social media, according to the Pew Research Center—relatively equal shares.[108] Are children displaying gender differences (beyond the realm of video games) that have consequences for STEM?
- More broadly, how will children's habituation to digital media affect their experience of the physical world, and their perception of STEM subjects?
- And, as Dean Emily L. Allen, Ph.D., observes, "We're all more distracted these days by social media. The ability to focus and concentrate is critical for learning STEM subjects. It's not clear yet if there are gender differences in the effect of social media on student performance in STEM subjects." Will differences emerge?
- As computer coding becomes a fundamental form of literacy, will habits of rigorous thinking motivate students to pursue other STEM courses?
- Initiatives to engage girls in STEM are attracting wide attention. Why not design and conduct a high-profile campaign to recruit adult women who want to change careers? What curricula and pathways could most easily lead them to STEM fields?
- In community colleges, how can efforts to strengthen and expand STEM curricula be accelerated? What's being learned from efforts to create smooth transfer pathways from community colleges to 4-year institutions with STEM degree programs?
- What opportunities for working in STEM fields do not require formal degrees? How can these opportunities be expanded?

- To what extent is what's perceived as gender bias in fact the result of a "supply problem," as suggested by research done by Cornell psychologists?[109]
- Peer review is not free from gender bias.[110] How is this best addressed? Liesl Folks, Ph.D., Dean of the School of Engineering and Applied Sciences at the University at Buffalo—SUNY, contends, "By educating students on their own gender bias." And she asks, "How is it possible that we have them at our institutions for 4 years, but there is every chance they will miss being educated on the role of bias and privilege while we have them?"
- How can more women be motivated[111]—and learn how[112]—to become STEM entrepreneurs?
- What strategies could rapidly increase the number of women seeking patents?
- Is the cybersecurity threat so severe that we need a national scholarship program to vastly expand our cybersecurity talent pool? If yes, who should design, fund, and manage it?
- Why shouldn't the NSF integrate gender goals into its global SAVI STEM crusade?[113]

And here's a closing question that's a happy one to dream on:

> The STEM gender gap exacts a steep opportunity cost: the time, effort, and talent now devoted to closing it are diverted from tackling other problems. When academia is finally attracting and graduating many more female STEM students, has created a far larger female STEM talent pool to recruit faculty from, and has removed barriers to their advancement, a valuable resource will be liberated for other pursuits.
>
> What should they be? What will they be?

References

[1] http://www.morganstanley.com/ideas/gender-diversity-investment-framework

[2] http://fortune.com/2015/11/17/stem-toys-girls-holidays/?xid=soc_socialflow_twitter_FORTUNE

[3] http://insight.ieeeusa.org/insight/content/careers/72521

[4] https://www.nsf.gov/statistics/2016/nsb20161/uploads/1/nsb20161.pdf

[5] http://www.bls.gov/opub/mlr/2013/article/occupational-employment-projections-to-2022.htm

[6] http://www.bls.gov/ooh/Computer-and-Information-Technology/Computer-and-information-research-scientists.htm
[7] http://www.bls.gov/ooh/Math/Mathematicians.htm
[8] http://www.bls.gov/ooh/computer-and-information-technology/information-security-analysts.htm
[9] https://www.naceweb.org/s09032014/starting-salaries-for-2014-graduates.aspx
[10] http://www.bls.gov/ooh/Architecture-and-Engineering/Civil-engineers.htm
[11] http://www.bls.gov/ooh/Architecture-and-Engineering/Petroleum-engineers.htm
[12] http://www.bls.gov/ooh/Architecture-and-Engineering/Nuclear-engineers.htm
[13] http://www.bls.gov/ooh/Architecture-and-Engineering/Aerospace-engineers.htm
[14] http://www.bls.gov/ooh/Architecture-and-Engineering/Computer-hardware-engineers.htm
[15] https://www.asee.org/papers-and-publications/publications/college-profiles
[16] https://www.asee.org/papers-and-publications/publications/14_11-47.pdf
[17] http://cra.org/
[18] http://cra.org/data/
[19] http://cra.org/wp-content/uploads/2016/05/2015-Taulbee-Survey.pdf
[20] http://web.mit.edu/facts/financial.html
[21] https://www.codecademy.com/
[22] http://www2.ed.gov/policy/rights/guid/ocr/sexoverview.html
[23] http://www.nsf.gov/statistics/2016/nsb20161/#/report/chapter-3/highlights
[24] http://www.aauw.org/research/barriers-and-bias/
[25] http://www.aauw.org/files/2016/03/BarriersBias-one-pager-nsa.pdf
[26] http://womenintheworkplace.com/ui/pdfs/Women_in_the_Workplace_2015.pdf?v=5
[27] http://scholar.harvard.edu/iris_bohnet/what-works
[28] http://www.wsj.com/articles/real-fixes-for-workplace-bias-1457713338
[29] http://www.majoracartergroup.com/
[30] http://bronxsoftware.org/
[31] https://www.weforum.org/agenda/2016/03/a-visual-history-of-gender-and-employment
[32] http://www.mckinsey.com/global-themes/women-matter

[33] http://www.mckinsey.com/global-themes/employment-and-growth/the-power-of-parity-advancing-womens-equality-in-the-united-states
[34] http://www.changetheequation.org/
[35] http://www.changetheequation.org/stemtistics
[36] http://blogs.edweek.org/edweek/curriculum/2014/01/girls_african_americans_and_hi.html
[37] http://home.cc.gatech.edu/ice-gt/556
[38] https://research.collegeboard.org/programs/ap/data/archived/2013
[39] http://createsend.com/t/y-C32A98B393E022E4
[40] https://nscresearchcenter.org/snapshotreport-degreeattainment15/
[41] https://girlswhocode.com/
[42] https://blog.cloudcalc.com/2016/02/24/happy-introduce-a-girl-to-engineering-day/
[43] https://tech.cornell.edu/impact/witny
[44] https://www.ncwit.org/sites/default/files/resources/womenintech_facts_fullreport_05132016.pdf
[45] https://www.nsf.gov/statistics/2016/nsb20161/uploads/1/nsb20161.pdf
[46] https://www.aauw.org/files/2013/02/Why-So-Few-Women-in-Science-Technology-Engineering-and-Mathematics.pdf
[47] https://www.technologyreview.com/s/600812/female-coders-are-more-competent-than-males-according-to-a-new-study/
[48] http://www.nber.org/papers/w17888.pdf
[49] https://www.technologyreview.com/s/601200/y-combinators-cofounder-talks-about-the-challenges-of-being-a-female-entrepreneur/
[50] http://fortune.com/2016/04/01/venture-capital-still-has-a-big-problem-with-women/
[51] http://www.theatlantic.com/business/archive/2014/08/women-are-more-likely-to-secure-kickstarter-funding-than-men/376081/
[52] https://www.whitehouse.gov/blog/2016/01/30/computer-science-all
[53] http://www.csnyc.org/computer-science-all
[54] http://www.nsf.gov/news/news_summ.jsp?cntn_id=127063
[55] http://www.nsf.gov/news/special_reports/savi/index.jsp
[56] http://www.firstinspires.org/about/vision-and-mission
[57] https://www.brookings.edu/events/stem-education-and-future-generations-of-american-inventors-technologists-and-explorers/
[58] https://www.c-span.org/video/?409222-1/nasa-administrator-charles-bolden-discusses-stem-education
[59] https://www.nae.edu/19582/Reports/24985.aspx
[60] https://www.ncwit.org/

[61] http://wtp.mit.edu/
[62] https://due.mit.edu/initiatives/k-12-outreach
[63] https://www.facebook.com/search/top/?q=pathways%20to%20cleaner%20production
[64] http://www.nyit.edu/events/introduce_a_girl_to_engineering_day
[65] http://www.nyit.edu/events/nyit_connect_to_tech_engineering_and_technology_high_school_showcase_compet
[66] https://www.scs.cmu.edu/outreach
[67] http://www.aauw.org/what-we-do/stem-education/
[68] http://www.huffingtonpost.com/news/girls-in-stem/
[69] https://girlswhocode.com/
[70] http://sheplusplus.com/
[71] https://twitter.com/hashtag/ILookLikeAnEngineer?src=hash&ref_src=twsrc%255Etfw%20v
[72] http://www.crainsnewyork.com/article/20160321/TECHNOLOGY/160319836/cuny-cornell-tech-launch-women-in-tech-initiative
[73] https://www.whitehouse.gov/administration/eop/ostp/women
[74] https://www.whitehouse.gov/issues/women
[75] https://www.asee.org/member-resources/councils-and-chapters/engineering-deans-council
[76] https://www.asee.org/documents/member-resources/edc/EDC-DiversityInitiativeLetterFinal.pdf
[77] http://societyofwomenengineers.swe.org/
[78] http://www.ieee.org/membership_services/membership/women/index.html
[79] http://cra.org/cra-w/mission/
[80] https://www.womenwhocode.com/
[81] https://code.org/about
[82] http://anitaborg.org/
[83] http://www.nytimes.com/2015/05/22/upshot/making-computer-science-more-inviting-a-look-at-what-works.html
[84] https://www.hmc.edu/about-hmc/mission-vision/
[85] http://www.cs.washington.edu/
[86] https://www.asee.org/public/conferences/1/papers/335/view
[87] https://tech.cornell.edu/impact/witny
[88] http://www.nsf.gov/discoveries/disc_summ.jsp?cntn_id=137951&WT.mc_id=USNSF_1
[89] https://en.wikipedia.org/wiki/Grace_Hopper

References 155

[90] http://www.huffingtonpost.com/monique-morrow/five-reasons-why-women-in-it_b_5198421.html
[91] https://www.intrepidmuseum.org/GOALSforGirls#1
[92] https://www.stevens.edu/news/stevens-awarded-nsf-grant-transform-stem-courses
[93] https://www.aacu.org/sites/default/files/files/publications/E-PKALSourcebook.pdf
[94] http://blog.cloudcalc.com/2016/02/24/happy-introduce-a-girl-to-engineering-day/comment-page-1/#comment-1425
[95] http://www.nytimes.com/2014/07/18/upshot/some-universities-crack-code-in-drawing-women-to-computer-science.html?_r=2
[96] http://qz.com/192071/how-one-college-went-from-10-female-computer-science-majors-to-40/
[97] http://mentornet.org
[98] http://www.nyit.edu/box/news/nyit_launches_entrepreneurship_and_technology_innovation_center
[99] http://www.wsj.com/articles/real-fixes-for-workplace-bias-1457713338
[100] https://textio.com/products/
[101] http://www.pnas.org/content/112/17/5360
[102] http://www.nsf.gov/funding/pgm_summ.jsp?pims_id=5383
[103] http://www.nytimes.com/2016/05/11/business/dealbook/puerto-ricos-fiscal-fiasco-is-harbinger-of-mainland-woes.html?src=me&_r=0
[104] http://www.nsf.gov/news/news_summ.jsp?cntn_id=127063
[105] http://www.luma.fi/news/2940/
[106] http://www.nsf.gov/statistics/2016/nsb20161/#/stateind
[107] http://engineeringeducationlist.pbworks.com/w/page/27610370/Engineering%20Education%20Research%20and%20Teaching%20Centers
[108] http://www.pewresearch.org/fact-tank/2015/08/28/men-catch-up-with-women-on-overall-social-media-use/
[109] http://www.pnas.org/content/112/17/5360
[110] http://www.sciencemag.org/news/2015/05/plos-one-ousts-reviewer-editor-after-sexist-peer-review-storm
[111] http://knowledge.wharton.upenn.edu/article/why-are-there-more-male-entrepreneurs-than-female-ones/
[112] http://athenacenter.barnard.edu/leadership-lab/course-catalog/entrepreneurship
[113] http://www.nsf.gov/news/special_reports/savi/index.jsp

6
Reshaping Traditional Ways of Learning

6.1 Unearthing the Magic of STEAM

By Shraddha Chaplot, Greengineer and Machinegineer (fun titles she created herself)

Shraddha Chaplot is a playful engineer who imagines offbeat ways to pursue her dreams, leading to her pastimes, passions, and pondering becoming her playground in life. Shraddha is also a huge enthusiast of being your authentic self, combining her love for engineering, STEAM, hands–on projects, and wacky ideas to inspire and empower anyone and everyone, and in doing so, enabling them to realize their true potential. She graduated from the University of California, San Diego with a bachelor's degree in Electrical Engineering and depth in Machine Intelligence.

I love numbers. I have played with them since I was a little girl, when the great Mathematical Wizard (my mother) introduced me to its wonders. She started me off with the magical world of times tables at the curious age of three.

We would sit together and she would say:

Math Wiz (aka my mom): Two one-za ...
Me: Two!
Math Wiz: Two two-za ...
Me: Four!
Math Wiz: Two three-za ...
Me: Six!

That was how my mom taught it to me, where "za" meant multiplying the two numbers stated before it and giving the answer.

I also remember writing my times tables up to 40, and for my mom, writing just the answer was never enough. I would have to write it out "$13 \times 7 = 91$". She believed that if you knew your times tables by heart, and that your mental math skills were solid, you could learn anything about, and do anything with, math.

My mother is truly a Mathematical Wizard. And she helped create one just like her.

I instantly fell in love with the beautiful patterns I discovered and created with the infinite combinations of these ten digits –> 0, 1, 2, 3, 4, 5, 6, 7, 8, and 9.

Because imagine this for a moment: every single thing we do that is based on numbers has something to do with these ten pioneers.

From building a spaceship to exploring stars, from the tools we use to construct homes to playing with our smartphones—the presence of these fundamental numbers is inescapable.

Look, I think we should just get this out of the way, right here, right now:

Yes, I am the girl who asked for more math homework.

Yes, I am the middle-school student who would stay up late into the night writing pages and pages of proofs and patterns for a simple set of math problems, color-coding, and adding a table of contents.

Yes, I am the college freshman who found mistakes in a professor's Calculus textbook and told him he should let me proofread (and solve!) all the math problems to ensure they were right. More than 1,000 pages! He let me. And I did.

Yes, I am the girl who still takes math tests online, for fun.

I am all of these girls and it is because I unearthed the beauty of math.

And I cannot even begin to tell you the beautiful numerical patterns I find in every single thing I see and do.

From calculating numbers on license plates to making patterns from daily dates, from math in the sky to the very popular π: the world is my wonder.

But this is not a typical story by far. People usually do not have the curiosity in math that I do, not even to a small degree. They find it to be nerdy, difficult, unnecessary, the "when am I ever going to use this?" culprit, and unfortunately—a prime reason for bullying. Because if you didn't already know, being smart usually is not considered cool.

I am here to change that perspective and introduce you to the unknown world of math, where you too can become a Mathematical Wizard! Or anything you want.

Enter: STEAM

Traditionally called STEM and based on the fields of science, technology, engineering, and mathematics, STEAM adds the very broad field of arts into the mix. And my love for math, engineering, and creating beautiful things is exactly what I envision STEAM to be about. But let's not be so focused on adding in art as if it were colorful sprinkles you optionally top your three-scoop waffle-coned ice-cream with. Art needs STEM and STEM needs art.

The truth is, art is as critical to experience as any of the other fields. The integration of STEM and art is key, and that leads to the illumination of our multiple facets.

What do I even mean by this? I could share countless things demonstrating the beauty of STEAM, but what I believe to be its most important underlying aspect is that it is each and every human embracing the many facets they have within themselves. It is through this that we can create incredible, unimaginable things—some big, some small, but each worth experimenting with.

Let me explain and provide an example.

I have never fully understood why people leave parts of who they are and their curiosities at home. The parts where they are fun or creative or have a hobby, oh heck, even have a sense of humor. (Did I just write "heck"?)

Why is it that, when we come to work, we are told that work is for work, and everything else is for our personal life? (That is a whole different conversation for another time.) I always felt that the more we brought our full selves to work—the passion, the things that drive us to create and build and be better humans—the more we could and would enhance the actual solutions we created.

So here's my example of STEAM, both in the literal sense and in my "be your multi-faceted self" sense.

Back in 2013, I was browsing Kickstarter, an online crowdfunding platform where people share their creative project ideas and get funding from those viewing it. I found one that I instantly knew was going to transform the cool factor of STEM—I felt like I had unearthed a hidden treasure!

This small start-up completely re-imagined the traveling circus by teaching kids about STEAM through interactive games of lasers, fire, and robots—oh, you BET I was going to be a part of it! I immediately backed the project and reached out to them. And what started as my personal passion project became something I eagerly wanted to bring into the company I work for—Cisco.

About two years later (mid-2015), I spoke with our Chief People Officer, Francine (Fran) Katsoudas, about partnering with the start-up for their San Francisco debut. I told her about how unique this event would be and how I envisioned what it could do not just for STEM, but also for this fairly new integration of STEAM.

Just like everyone I had spoken to for the last two years, Fran had never heard of this traveling circus, but after seeing what I envisioned, she was convinced and came on board.

We ended up becoming co-sponsors of the main extravaganza when they came to San Francisco in November 2015. The experience created for that weekend was incomparable and unforgettable. Fran later told me how cool her children thought Cisco was after she took them to the traveling STEAM circus, and that it was because of this grand vision I had had!

You see, my plan worked—those kids would now associate cool with Cisco and cool with smart! Just like STEM, I love showing people the cool factor of things they never thought. It's there—you just need to be exposed to it. And guess what—everything I did for Cisco to partner with this start-up had nothing to do with my "job description." I did it straight from my personal passion and brought it into the company I worked for. I made it happen.

The reason it worked was that I brought in all of me—I brought in an external passion (STEAM), I brought in my external friends from the start-up, and I created this completely new experience to show the power of an individual (I'm a Greengineer and individual contributor at Cisco). I brought in my creative, educational, fun, and playful facets.

For me, THAT is STEAM.

STEAM is about embracing the many things we all are. By doing so, I was able to create and realize an incredible idea I had.

And when you do that, the magic is inevitable.

So how can we continue this magic? How can we get more kids (or anyone!) excited about STEAM? What can we do next?

- First, dare to be curious: play and explore things. Break things, make things, and rebuild them. I became an engineer because, as a little girl, I loved taking things apart to see how they worked, or sometimes, just to break them. And I was always fixing things, which is why my parents called me their "Little Fixer." If something is destined for the trash, like an old electronic device or even a teddy bear, take it apart and see the guts of it. Research it. Find out how it is made, what materials are used, and how it was put together. Ask every question you want of it and try to find the answers.

- Give young children role models to look up to—ones who they can relate to and ones with what I equate to this formula: Magic Mix = Good Heart + Curious Brain.
- If someone does not like technology or anything STEM-related, do not force it. Not all of us have to be techies or engineers or physicists. That's the beauty of humans—we are so diverse and that should be embraced. Let everyone explore what they love or do not love to do, and just give them opportunities to try out and experiment with different things.
- Paint. Draw. Build. Sketch. Play. Dance. Anything. Even if you don't know how. *Especially* if you don't know how. You don't have to be perfect or the best at it. Don't deny it. Try it. And what you will find is that each of these things has so many facets of STEM.
- Don't just code. Research. Visit facilities that have free tours. You never know what might spark your interest (recycling centers to see how waste is re-used; local astronomical societies to gaze at the sky even if you can't afford the equipment; museums, windmills, and so on). The power of doing things you would not normally do is what sparks new-found interest and lifelong passions.
- Recognize that, no matter what, everything has an element of STEM. Everything.

I believe that each of us is born with many unique elements of creativity and curiosity. And for some reason, as we get older, we are taught to be less creative and curious about the world around us. That is why it is so important for us to show children and youngsters that these are attributes to be embraced as we get older, not to be banished into the forgotten depths of our brains, never to be seen again. This creative and curious spirit is what will help children become young adults who find unique solutions to solve so many of the challenges we face today, as well as create new things we never dreamt of. They will not come up with the same solution as another person. They will connect things in ways that seem impossible, and because of their exposure to STEAM, they will make it possible. STEAM will show kids how to be true to the many amazing facets that make them up.

What are you waiting for?! Be like February 8, 2016 (2816), and hop on this STEAM train! You'll be in for the most magical ride of your BrainHeart's life! (You can easily find me, as I am the engineer welcoming you aboard.)

Follow Shraddha on Twitter, take a peek into her playful brain, and join in on her spontaneous adventures!

6.2 Changing University Education through the Internet of Things

By Rahilla Zafar

Sandra Bradley is the Research Director of the Internet of Things (IoT) Lab at the University of Wisconsin-Madison. The IoT Lab is a campus program that enables hands-on experimentation and promotes entrepreneurship.

Sandra has always been fascinated by how people use technology to communicate and interact. "I'm interested in what types of tools need to be developed for people to accomplish their goals," she says. "That's been a common thread throughout my career."

Her academic work and early professional career focused on communications and marketing. In her university position, she works with many Midwest-based companies to help them better leverage technology and understand the digital marketing age and information technology (IT). During these experiences, she began to hear a lot about peer-to-peer learning and collaborative education. "We got at what works and what doesn't, but we didn't get at, 'What's next?'" she says.

The Internet of Things Lab at the University of Wisconsin-Madison gives executives access to talent and pipeline development from a team of faculty researchers and tech-savvy students. "We could really provide value for students with hands-on experience, real-world problems and connect them to companies where they could learn," she says. "It's a reverse mentorship model."

The lab has received a great deal of attention from the media and companies. Still, it got started in a more humble "boot-strap" fashion, as Bradley describes it. The faculty found used equipment to stock the lab with, and then invited students to experiment with it. That dynamic creates an environment Bradley likens to a "sandbox."

"We call it 'sandboxing.' We invited students to come in and experiment with faculty in a shared space," she says, adding, "There was tremendous interest." The lab holds a showcase every semester, so the teams had an incentive to develop the best ideas. "Students loved the idea of being able to work on their own project," she says. "We provide the roadmap and some framework and tools and resources for them, as well as mentorship and education. The students all worked with no compensation or academic credit. They came because of their passion."

"The spirit of exploration, passion, and innovation fostered in the lab goes nicely with the goals of a four-year degree program because it offers a hands-on supplement to their studies," Bradley says.

The lab has seen participation from students studying such diverse fields as philosophy, media communications, business, engineering, computer science and physics.

"In a traditional model of education, many students don't have the opportunity to engage with students from other departments," says Bradley. "The lab provided an exciting place to do that. The ideas generated by multi-faceted teams were exponentially better."

Bradley's role has developed from getting the IoT Lab off the ground to mentoring students and helping manage the intellectual property (IP) process. The lab can even boast of having two teams go into an accelerator program in the area. "Those teams are really taking that next step," Bradley says. "While the lab is focused on early ideation to help build a business case and grow, they are now in an accelerator program, which is a great next step for them."

Today, the concept of the four-year degree has evolved considerably from what it was just a few decades ago. The focus now is more on building skill sets to better prepare students for a highly competitive landscape. The teamwork from students of all different majors in the IoT Lab answers a key question in education: How do educators best prepare students to understand the scope and challenges of working in a multidisciplinary team?

Promoting Women's Interest in Engineering and IT

In the IoT Lab's first semester, 200 students submitted proposals, of which 50 were accepted. When Bradley looked around the room, she noticed a huge gender disparity: 47 students were men, and only three were women.

"I thought, this seems really skewed," she said. "I looked deeper at the College of Engineering, and I was really dismayed to find such low participation from women. I thought, 'There must be something else we can do.'" Nationally, only 14% of engineering students are women.

Bradley approached the Society of Women Engineers, which has a strong presence in Madison, and encouraged them to be part of the lab. She worked with the national chapter president, who served as an important role model, thus helping bring more visibility to women in engineering.

Bradley also worked with the IoT World Forum and Cisco's Young Women Innovation Challenge. She established a series of "bootcamp experiences" that invited middle-school and high-school girls to come in and play in the lab, learn what IoT is, and discuss its challenges. In all, Bradley led four of these bootcamps, each of which had about 20 girls.

"This age range is so enthusiastic," says Bradley. "There is so much opportunity to raise awareness about the options they have and to build their confidence. I find that even college-age women look around at these heavily male environments that are not always welcoming and open."

Each boot camp attracted local women in engineering, who came in and shared their experiences with the girls. The Society of Women Engineers sent a team to facilitate each such session and talk about their passions.

"Their involvement," she says, "is quite inspiring."

The lab is also used as a meeting space for the university's Women in Tech Group. "I'm hoping we can make the lab more attractive to women and create a more welcoming environment. I don't want it to be perceived as some inaccessible place." Bradley has started to see more female students in the lab's information sessions.

"Men and women do think differently. From my observations, women can put together more unrelated things to make sense of the result," says Bradley. She describes a project where a male student wanted to develop a medication reminder band for seniors. The student knew how to build it, but wasn't able to think through all the other considerations. One female student with a retail background was able to take all the different points of the bracelet and think about how they answer key questions. She even suggested that they consider the bracelet for other age ranges, such as kids going away to camp.

"We have engineers who can build. But we need them to understand whether there is a market for their idea so people can use it. All of these things are important. We want to elevate the lab so that it's for people who want to solve problems, and not just those who like to tinker with stuff," Bradley explains. Women, she says, really understand how to create a collaborative, highly communicative team. They also know very well how to work across disciplines.

"That is what we are driving toward," Bradley says. "We want students to have the ability to think flexibly with a diverse team. We encourage multi-faceted thinking with a linear and non-linear perspective on how technology can work." Non-diverse teams often get stuck and don't know where to go," she adds. She encourages students to push deeper instead of doing the same thing over and over again, or relying on things that didn't work in the past.

Achieving Directional Progress for Women

Bradley is a big advocate of STEAM over STEM. As we have seen, the latter stands for science, technology, engineering, and math. The first term, as we have also seen, adds 'arts' into the mix. "Creativity," Bradley says, "must be valued as a core component of IT and engineering, and may even help attract more girls to STEAM at an early age."

Early middle school is a good time to expose girls to topics such as engineering. As an educator, Bradley believes there is a big opportunity for middle-school science teachers here. But she says not many teachers know how to cultivate, inspire, or encourage. "There is so much work to be done in this area. Once girls are in high school, they've often made up their mind on whether they're good at something," she explains.

In order to attract girls to STEAM, Bradley says, there needs to be a social or peer-to-peer component. "So many of the girls I know need to be connected and feel as though their peers are interested, too, and that what they are doing has value. They don't want to be singled out as being weird."

"There also needs to be awareness of engineering as a field," she says. Young people in their daily lives see firemen, teachers, and so on. Here there is a cultural component, and a need to send out the message that tech isn't an exclusive field—that everyone is welcome.

"What amazing problems you can solve with technology," says Bradley. "But change has to come from a number of different places. First, education needs to change, with a total reformation, and an awareness of and a drive for girls in elementary and middle school, especially around STEAM."

6.3 Education and "Being a Girl"

By Ankur Kumar

Ankur Kumar received her undergraduate and MBA degrees from The Wharton School, where she also served as Director of MBA Admissions. During her tenure at Wharton, she led efforts to increase the gender and industry diversity of the student body. Under her tenure, Wharton became the first leading United States MBA program to achieve a student body comprised of 40% women. She also helped increase entrepreneurialism in the class body — students she admitted went on to found companies across all sectors including CommonBond, Omaze, Skillbridge, Stylitics, and Senvol amongst others. Ankur is now a leader in the digital-education arena.

My parents came to the United States from India in the mid-1970s to further their careers—my mother is a physician; my father, a retired engineer. In my house, education was the ticket out. As I was growing up, my father would often tell the story of how his family had fled Lahore during the India-Pakistan partition. Their family was on the wrong side of the subcontinent at the time, and they lost everything when they fled to Delhi. The story always came back to one fine point—that my grandfather was able to rebuild his business and a life for his family because he was educated. My mother's family also valued education. In a family that includes six sisters, all were educated. Her own mother, though she never worked outside the house, held an undergraduate degree—a rarity for Indian women of that generation.

When I was growing up, education was not "negotiable," not a "nice thing to have"—it was a responsibility, an opportunity, and a gift. It was what my parents cared most about for me (health and happiness aside, of course). As I look back, it's no surprise that I found my calling in opening up access to education to others. My own educational experiences opened up my eyes— even as they also opened doors. And enabling others to experience that is what ultimately motivates and fulfills me.

I also grew up in a household where my gender never defined me. My parents never treated me differently because I was a girl. It was not uncommon for their generation to deprioritize a daughter's education. My father would proclaim proudly that he never treated me differently because I was a girl. They prioritized my education, sent me to private school, encouraged me in math when I took to the subject early on in school, and encouraged learning both inside and outside the classroom.

My mother was more accomplished professionally than my father—the result of a conscious choice my parents had made to advance her career. My father was a supporter not only in words, but in action. It was my father who picked me up from school, drove me to dance class, and came to my track meets. He did the grocery shopping, and to this day is the best at ironing and vacuuming in our family.

I recognize that I was incredibly lucky. Not only because of the access I had to a high-quality education—but because I grew up in a household where I experienced firsthand a strong, professional mother and an encouraging father. Many girls never receive that kind of encouragement from their families, teachers, or community. They aren't told they can do or be anything they want. Even worse, many girls are told the opposite: that there are things they can't do or professions they can't pursue.

My mother showed me, by example, that I could—and should—pursue a career that had meaning in my life; that it was not only OK, but absolutely vital, to do so. She showed me that life was and would always be a balancing act, a matter of give and take, and that the best partner was one who would support that balance.

It never occurred to me that my family and experiences were unusual (for the times or the community). In many respects, I'm grateful for my naïveté. It's because of it, and because I had my mother's example, that I had the freedom to figure out my path and the confidence to pursue it (not without doubts and help along the way, of course).

Eyes Opened

It was only when I ventured out into the working world that I started to see that it did not line up with what I had experienced in my bubble. It was the first time in my life that being a woman was something I was acutely aware of—and not in a good way. I was recruited into a coveted group—Technology Mergers and Acquisitions—at a prestigious investment bank. I remember calling my mother excitedly to tell her that they wanted me! While I didn't have a technology background, I had been a finance major in college and had the confidence to be a fast study. And while I was lucky to be surrounded by an affable group of colleagues and supporters—with whom I'm still in touch to this day—I can't tell you the number of men who commented that I got the spot in the group only because I was a woman. I let that roll off, fortunately. One thing I learned from a tough, stoic mother, was to develop a thick skin.

At the time I don't think I could have put my finger on it. But in retrospect—and while it wasn't explicitly said—it felt like there was an understanding that as a woman you had to work harder to prove yourself, be even better than your male peers. Any sign of being a "woman" (whatever that meant) put you at risk of devaluing or discrediting your capabilities. I look back and realize that in those years it was almost as if I tried to cover up and hide being female. There were also few aspirational women role models in my immediate environment.

It was also during this time that I began to get more exposed to technology. It was still the early days of the Internet, and I had a window seat, given my industry focus. I would pore over primers on semi-conductors and routers. Even though my father was an engineer, technology always felt a bit foreign to me. I struggled with physics, not grasping circuits and series as fast as my dad tried to teach me. I was honestly a bit intimidated by technology, now that

I think about it. And because it didn't come so naturally, I wasn't as motivated to invest in learning about it.

Later in my career, when I began to get more involved in the operational side of business, technology and systems became a daily part of my life. I spend my time now in the digital learning space. Despite wanting to be on the forefront of where education is heading, I was initially intimidated. I felt almost like a fraud, having more experience in the traditional education space. I don't know how to code, I still struggle with vectors, but I am less intimidated after having broken down my mental barrier. Instead of focusing on what I wasn't—an engineer, a technologist—I focused on what I was: curious about what technology could enable. To do this, I started to read more and ask questions to help me keep up with the trends in my space (talking to the engineers on our team helped a lot!).

The Wharton Years

When I joined the Wharton MBA Admissions team, the value of the MBA was being heavily questioned. Anyone who knows me, knows I don't like the hard sell—I'm not an arm twister. But I do believe in the power of information, access, and options. My own business-school experience had been a kaleidoscopic journey of self-discovery and learning. I had decided to go back to school to focus on studying people and strategy—my intention had been to work in corporate talent or as an executive coach. I never once stopped to think that my choice of profession made an MBA any less meaningful or relevant. The opportunity to learn about all parts of a business, to get to meet and learn from peers with diverse professional and personal experiences, and to try (and fail) in a lower-risk environment all resonated with me.

However, I faced misperceptions by classmates who thought that pursuing such a field was less "valid" than another post-MBA path. It was my bullish confidence that allowed me to take my classmates' perceptions as a challenge. It was my mission to help change their view—to help them realize that this was a topic of importance no matter what career path one pursued. It was a coup when I was able to bring together several classmates (including several men!) to start a Human Capital Management student club with me.

It was this mindset—of carving out my own path and helping change perceptions—that I brought to the Wharton MBA Admissions team when I joined in 2009. TARP funding had put severe restrictions on financial services firms' hiring at the time. Since they had been among the largest hirers up to then, this had a devastating effect on the number of "secure" opportunities many graduates came to business school seeking.

It would be easy to say that increased interest in entrepreneurship was a back-up, a fallout from the financial crisis. But that wasn't the case. After the number of "steady" professional opportunities dwindled, the crisis forced many students to take a step back and really reflect on what they wanted to do. What emerged was a new opportunity. Many people wanted to create something of their own, or to build a business from the ground up, or disrupt an industry. However, there was still a misperception that business school only made sense for careers in certain industries. Entrepreneurship and technology were not sectors that came to mind when one thought of an MBA. However, I knew from my own experience that this simply wasn't true.

I started to engage students, alumni, and employers in a dialogue to try and change the equation. When I was a student at Wharton, I could count on two hands the number of "entrepreneurs" who were in my graduating class. By the time I left Wharton MBA Admissions, that number had grown tenfold. And over 10% of the 800+ incoming class had founded or worked at a start-up pre-Wharton. Similarly, the number of students who went into technology internships between their first and second year crossed the triple-digit mark and rivaled the numbers joining more incumbent MBA recruiters, such as financial services. Getting here was not about the numbers or about having stats to publish. It was about a deep appreciation for diversity of all kinds. Whether professional, personal experience, gender, or leadership style, innovation comes from bringing together people who think and approach situations differently. By expanding the diversity of our student body, we were helping to expand minds and to create an opportunity for innovation that would impact individual lives, but also industries and even countries. I'm proud to point to start-ups from my tenure, such as CommonBond, which is revolutionizing the student lending space, and to Skillbridge, which is revolutionizing the talent space.

At the time, there was also increased discussion about women in business. Sheryl Sandberg and Marissa Meyer reigned supreme. As I look back, I see that this conversation was personal. As a woman in business, who had the good fortune to be recognized and rewarded—but who also struggled with not having women to look up to, I wanted to change the experience for those coming after me.

I sought to help change the conversation. Working closely with our students, alumnae and other groups, we started to have a more intimate conversation with women. We wanted to better understand what they needed and to help them understand how an MBA could help them. We broke down misperceptions—of a lack of women in the MBA community, of a cut-throat culture, of a lack of opportunities in a wide range of industries—to help make women aware that this was a community that they could truly flourish in. We also took the opportunity to engage our alumnae. We were fortunate to have many successful women whose stories needed to be told and heard.

One of my proudest moments—and legacies—was not when we first crossed the symbolic threshold and had 40% women in the program, but when we did it again, and again and again. While we still have work to do in schools and in the workforce, we were able to create a space for women and men to engage in constructive dialogue. Wharton now has a group of men—the 22s (named after the percentage wage gap between women and men)—who are advocates, cheerleaders, and supporters of our women.

I remember speaking at an event a few years ago to a group of women who had been admitted. I recounted to them how being a woman was something you used to have to dance around, and how talking about balancing work and family was considered a weakness or grounds to question your commitment. It makes me beam to know that only half a generation later, thanks to the efforts of many women and men, we have changed the conversation. And to have been a part of it is one of my proudest legacies. That being said, we still have plenty of work to do. We need to start the conversation even earlier with women—to ensure that from an early age they are ingrained with the notion that they can be or do whatever they choose. And we need to create more opportunities and support during school and in the working world.

6.4 Creating a Culture of Peace and Understanding through Technology Education

Zika Abzuk is a Senior Business Development Manager at Cisco Israel, where she is leading Cisco's new initiative, Country Digitization Acceleration (CDA). Zika founded The MaanTech Program, which supports the integration of the Arab community in Israel into the high-tech industry under the auspices of President Shimon Peres. Together with Moshe Friedman, she also founded the KamaTech Program, which supports the integration of Ultra-Orthodox Jews into the high-tech industry. Prior to leading CDA at Cisco Israel, Zika led Cisco's Corporate Social Responsibility efforts in Eastern Europe, Israel, Palestine, and sub-Saharan Africa. In this role, she managed the implementation of several social investments, including a $10 million commitment in five African countries, and a further $10 million commitment in the Palestinian Territories to support the development of the high-tech sector. Thanks to these *efforts, the Palestinian high-tech sector went from virtually zero to creating 6% of GDP. Across many countries in the Arab world, enrollment in Cisco's Networking Academy is up to 64% female. However, in Israel, only 12% of those enrolled are women. Zika's efforts in bringing the Networking Academy to Arab communities in Israel and Palestine supported Arab youth in having technology exposure and access they had not had.*

What follows is an edited transcript of an interview conducted by Rahilla Zafar.

During the seven years in which I lived in the United States, there was always a key question on my mind: whether to stay in the United States or go back home to Israel.

My family and I decided to move back in 1999 after watching Prime Minister Yitzhak Rabin and then Minister Shimon Peres shake hands with Palestine Liberation Organization Chairman Yasser Arafat on the lawn of the White House. To me it represented a moment when we knew we wanted to be part of this new Middle East.

By the time we arrived in Israel, Rabin had been assassinated. I worried because I felt I had taken my kids from a privileged life in the United States to a place with no hope. That was depressing and a difficult time for us. We also found that Israeli society had become more polarized, which was painful.

Connecting Israeli and Palestinian Youth through Tech

At Cisco, my first role involved managing a research-and-development team. One day John Morgridge, who was Chairman of Cisco at the time, came to visit. There were hundreds of Cisco employees in Israel, and he made a point of speaking to each and every one of us.

When I met him, I told him how I came to return to Israel and about my desire to foster a more cohesive society in Israel, through working with youth from different communities. He recommended that I check out and possibly utilize the Cisco Network Academy Program. That gave me the idea of bringing Jewish and Arab students together to learn and to know each other through technology. By then, the Israeli high-tech sector was already considered a big part of the country's success. I knew that it would be tempting for teenagers to be part of that.

By 2000, I started doing volunteer work to bring the Network Academy to Israel and Palestine. It helped me forge friendships with Palestinians in universities in Palestine. Not knowing much, I had started going to universities in the occupied territories and found amazing partners to work with.

They were very excited about the Network Academy. But at the beginning, they were also suspicious. It took a year to build trust with the universities and for them to really believe that I wanted to bring the academy program to their universities, colleges and schools.

A key obstacle for me was the lack of equipment—labs, routers and switches—because the universities could not afford to purchase them. I started looking into options for collecting donations. I learned that Marc Benioff, the founder, chairman, and CEO of SalesForce, wanted to donate to the Networking Academy and that he was very enthusiastic about Israel. He agreed to donate the equipment needs for universities and colleges in Palestine as well.

When John Morgridge advised me to check out the Network Academy, he promised that if I made it happen, he would attend the program's launching event. We had 10 academies in Palestine and 12 in Israel. I asked him to come to the launch of both.

I still had some funding from Benioff, so I decided to create a class for teenage girls in IT. I had enough money to get a teacher and run it as an after-school program. So I went to Nazareth (an Arab city in Israel) and Nazareth Illit, (a Jewish city nearby). Almost immediately, I had 24 girls from both the Arab and Jewish cities who wanted to learn technology.

After a while, the girls wanted to open the class up to boys. They felt they needed to include boys for it to succeed. We selected 50 students—25 girls and 25 boys, half Arabs and half Jews—and we mixed them to create two classes.

It was a very successful program. Orgad Lootski, a wonderful teacher, assigned them projects to do in small groups, which required them to visit each other's homes to complete the work. This arrangement helped create friendships. It was the first project ever to require those kids to work together and practice being on the same team. A lot of the prejudice melted away.

All 50 students studied together for a year and a half. Once a quarter, I would come and interview the students on how they felt about the program. Afterward, we would do something social such as bowling. We learned from them that they really felt empowered by the program and learned a great deal, but they wanted more social activities. This was before the Intifada started.

The Intifada started six months into the program. The students continued coming but they all felt they had to be very careful in what they would say as these were politically tense times. They shared that they would have liked to talk about everything and bring every bit of who they are into the program without being afraid. They wanted to discuss societal issues, and not just technology.

That's what gave me the idea to start a youth program across the country that would bring young people together for social activities. It was a youth movement, really.

First, it brought Israeli and Palestinian teachers together. Next, we brought in the youth, starting with Arabs and Jews within Israel and then extending it to Palestinians living in the occupied territories. This program has been running for 12 or 13 years. Today, it has seen 5,000 graduates. There are about 1,200 current students, and it has become a real success in Israel and Palestine.

I also extended the program across the Mediterranean, using different social networks to bring kids together across the entire region. Participating countries included Israel, Palestine, Morocco, Egypt, Yemen, Turkey, Cyprus and Portugal. We recruited young people in their late 20s and early 30s who had participated themselves to help us run the program. During the summer vacation, they met in their own countries and worked with the youth on various projects, and collaborated online to bring them together.

Investing in the Palestinian ICT Sector

John Chambers, Cisco's CEO at the time and now Chairman, was asked by President George Bush to lead a delegation of business people to rebuild the south of Lebanon after the war in 2006. John made a commitment that he would invest $10 million in the country.

At the time, John had not visited Israel yet. The north of Israel suffered from the war as well, so John decided to also make a commitment to invest $2 million in Israel. Around that time, Shimon Peres became president of Israel. As a minister, he had really liked the work we did with Israeli and Palestinian youth.

We had this great idea for President Peres to invite John to launch the digital-cities initiative we had worked on, and that would connect Arab and Jewish cities. We held a video-conferencing event where John met all of the instructors in the participating countries. We also celebrated both the new program in Nazareth and the Mediterranean program, which is called MYTech.

When John arrived, I went with him to Ramallah, a Palestinian city in the West Bank, where we met with President Mahmoud Abbas. John made a commitment of $10 million to invest in what he called a new model of job creation. Our Palestinian friends were very happy about it.

We selected one representative each from the president's office, the government, the private sector and the Palestinian IT Association. This group became our advisory board and served as our eyes and ears in Palestine. They said they had watched with awe as Israel had become a start-up nation and a high-tech superpower. They also said Israel and Palestine are similar in many ways. They are both small countries that don't have a lot of natural resources. The best resource for both is their people—entrepreneurial people who are highly educated.

Palestinians are among the most educated people in the Middle East. They wanted us to help them achieve the same type of high-tech industry as Israel has. It was an amazing challenge. I started to learn more about how the Israeli government created a successful tech sector. In Israel, Cisco had around 700 people, most of whom worked in research and development. I asked some of my colleagues to come with me to Palestine to learn the ecosystem, understand what would be required, and see what was already in place.

They agreed to come with me to Ramallah, and I set a date. A day before the visit, they approached me and said, "We're not going to tell our wives we're going to the West Bank because this is risky and dangerous. We're putting our lives in your hands, so it's your responsibility if anything happens to us."

These were men who had served in the army. I called my Palestinian friends and asked what I should do. I didn't want to take responsibility for their lives, so I suggested we meet in Jericho, at a hotel about 100 meters from the checkpoint. If something happens, I figured, they could run to the checkpoint.

When we arrived at the checkpoint, we learned it was the day before Independence Day in Israel, so there was a closure in the West Bank. No Israelis could come in and no Palestinians could go out. We grew concerned we had come a long way for nothing. A soldier at the checkpoint who had been watching us with interest pointed out a gas station about 500 meters away that was in "no man's land." Both Israelis and Palestinians could go there without a permit.

We met there, underneath a Bedouin tent at a gas station, with the leaders of Palestine's ICT sector. We met amazing entrepreneurs that day from Palestine. It was right after the elections in the West Bank where Hamas had won, resulting in the Palestinian economy almost coming to a halt. We met entrepreneurs who owned small companies with 15 to 20 employees who were working on several local projects. A few of the CEOs had studied in universities in the United States and Europe. They told us, "We know Cisco is such an important company and we want you to help us change the image of Palestine as a place that's open for business."

I suggested we should do it ourselves, with Cisco starting to outsource work to Palestinian companies. That was really a worthy challenge. Our team had been outsourcing work to India, which was a few time zones away and had a completely different culture. The advantages of working in Palestine included being in the same time zone, sharing a similar culture, and even a similar accent (many Palestinians speak Hebrew). We thought it could be a good start, but we had to send it to our research-and-development department.

I went to Tae Yoo, head of Corporate Social Responsibility for Cisco, and asked whether we could move forward if we funded the work. They agreed that Cisco would fund it for the first year from John Chambers' $10 million commitment. If they were happy with the Palestinian work, they would continue to employ them and pick up the cost.

We started with 12 engineers. The research-and-development team interviewed around 20 companies, and we selected three. Each business unit selected a different company and started outsourcing work to them. A year later, 35 Palestinians were working for Cisco, and the research-and-development team picked up the cost and are still employing them. The Palestinians are now part of the core team. It's a great success story.

Building a New Image for Palestine

We promised to change the image of Palestine, as a place that's open for business, so we started knocking on the doors of multinationals. We went to Microsoft, Google, HP, and Intel. Together with our Palestinian partners, we shared with them our success story.

Dave Harden of USAID offered to fund the initiation of these projects, which allowed Microsoft and others to get started outsourcing work to Palestinian companies as well. We eventually created over 1,000 new jobs in Palestine. I also worked with Yadin Kaufman and Saed Nashef to help start Sadara, the first venture-capital fund in Palestine.

We also developed a capacity-building program for the companies that had not been selected for outsourcing opportunities. We worked with people from the high-tech industry in Israel to conduct workshops. The high-tech ecosystem is like a contagious disease. When you infect others with it, you can open up to the world. It's something you can teach, introduce, and make happen in other places.

176 *Reshaping Traditional Ways of Learning*

As Israelis, we are not allowed to enter Gaza in the way we can go to the West Bank. As a result, it's more difficult for us to support Gaza's high-tech sector from Israel. This is part of the reason why Google partnered with MercyCorps to launch the first accelerator there, Gaza Sky Geeks.

When we started working in Palestine, we were hopeful for political change. Today, the situation is very depressing. You need courageous leaders who can make the change or do the type of work we've done. If we have more momentum, more people will demand that their leadership work harder for peace negotiations.

I'm optimistic, because I know people on both sides. We've had great partners to work with, who became our friends. Even though the situation is political, it is up to us to change it, and eventually the governments will follow. I truly believe in this: the Palestinians are entrepreneurial in spirit and highly educated. They can develop their own ecosystem for the digital world if they are empowered to do so.

6.5 Internet of Women Pakistan: Divided by Access or Skills?

By Iffat Gill, Founder and CEO of ChunriChoupaal - The Code To Change

Iffat Gill is an international NGO leader-activist, digital strategist, and social entrepreneur working on gender equality and economic empowerment of women through digital inclusion. She has been at the forefront of shaping the global policy debate around digital inclusion of women. She started initiatives like "Work To Equality" and "The Code To Change" to empower women to participate in the global digital economy.

Acknowledgement: I would like to thank Arno Meulenkamp for patiently editing the first drafts of my work for this book, and Diana Eggleston for painstakingly proof-reading.

I founded ChunriChoupaal and The Code to Change, a leading initiative bridging the digital gender divide by addressing the e-skills gap. My journey to empowering women started with my human-rights work in Pakistan, which was a way to help those without access to a high-quality education and to help women join an industry that provides most of the new jobs.

When I was 13, I got access to my first computer with an Internet connection. We were living on the sunny Maltese islands in the Mediterranean in those days. It was the days of noisy modems connecting over telephone lines. My father was very supportive of embracing new technologies. In high school, I wanted to study computer science. My family is of Pakistani origin, and in sub-continental communities children are "encouraged" to study either medicine or engineering.

I went to my school head and shared my interest in computer science in addition to my majors, which were biology and chemistry (I was not considered a future engineer, being a girl). Surprisingly, he made an exception. I could miss one biology class per week to attend a computer-science class, where, as it turned out, I was the only girl. Even though the teacher and the male classmates were very nice and helpful, I felt out of place. I often noticed that, when I entered the classroom, everyone would go quiet. This discouraged me and others from asking questions and it didn't allow me to learn and grow. I finished the coursework and passed the exam, but dropped the idea of continuing any further with it.

After high school, I moved to Pakistan at the age of 17 and continued studying pharmacy. Moving to Pakistan was a big change for me. Malta had had compulsory secondary education since 1971, and had an adult literacy rate of 99.5%.[1] The literacy rate in Pakistan varies regionally, and by gender. Pakistan's average literacy rate has been fairly steady at 58%, according to the Pakistan Social and Living Standards Measurement (PSLM). Girls' education is not a priority in rural and suburban areas of the country, especially among the less privileged demographics. Girls are actively discouraged from going to school. My family lived in the outskirts of Multan City, which is home to around three million people. Girls in the neighborhood were wondering where I went every day. When I told them I went to university, they asked me why I was studying. Surely I could learn to cook nice meals for my future in-laws without a university degree. Many who asked me these questions were women my own age, who had never had the opportunity to get an education. These girls were curious, bright, and hungry for knowledge. These questions and conversations led to my work for human rights and women's empowerment in Pakistan. I combined my university studies with volunteering for the Human Rights Commission of Pakistan and at a skills center that taught new skills to girls from the neighborhood.

As time went on, I became more connected with local human-rights and women's empowerment organizations. My network helped me implement different empowerment projects for girls and women. The vibrant civil society of Multan was at the forefront of challenging the laws and customary practices that discriminated against women. I was one of the youngest female activists who spoke against the Hudood Ordinances, a set of laws where a woman could be the victim of rape, or even gang rape. The law required victims to produce four male witnesses to the assault to speak on her behalf, and if the woman pressed charges without the strict requirements for evidence, the authorities would then assume the sexual act to be voluntary and the woman would be accused of *zina* (adultery) and incarcerated, unable to prove the crime. The law was put in place in 1971 by military dictator Zia-Ul-Haq's Islamization process and was extensively revised in 2006 by the Women's Protection Bill under President Pervez Mussharaf.

This wasn't the only issue we were trying to change. Forced marriages, acid attacks on women, and honor killings were also issues on which I, along with many others, were active. A Pakistani filmmaker, Sharmeen Obaid Chinoy, made two documentaries on some of these issues, each of which had won her an Oscar. Chinoy's documentaries, "Saving Face" and "A Girl in a River: The Price of Forgiveness," amplified the voice of the civil society that has been fighting against these atrocities for over two decades now.

Chinoy's films showed that media—not just the medium of film, but social media and the Internet—is a powerful way to get attention for a topic.

Speaking at an event against violence against women in Multan Pakistan, organized by Human Rights Commission and Mukhtar Mai Women's Organization, circa 2005.

6.5 Internet of Women Pakistan: Divided by Access or Skills?

It was said that after civil society started to get involved with issues such as rape and honor killings, the number of incidences of these events actually went up. What wasn't clear initially was that not only was the message communicated that something could be done only by speaking up, people also had an opportunity to do so by calling for help when they got in trouble. Mobile phones changed the entire society. Rashid Rehman, one of the most vocal human-rights advocate of the Multan region once stated that mobile phones have made justice accessible to women who previously did not know how to connect to get support and help.

Mobile phones and the Internet were not all good news for women, though. Free social interaction with the opposite sex is still a taboo in Pakistani society. But the behavior of young people changed significantly over the past decade with the advent of modern tools of communication. While tools such as mobile phones empowered people, they also posed security and safety risks for young women, leaving them vulnerable to exploitation and violence. Boys will routinely call or send SMS messages to random phone numbers with eloquent messages such as "I want to Friendship you" [sic].

I was always interested in finding ways to use technology to improve lives. So I took up specialized courses in Internet governance online. I got involved with international networks and like-minded individuals who were using technology to bring change to their communities. They also discussed the emerging issues that were posed by Internet access in different cultural and political contexts around the world. My Internet connection and digital literacy had empowered me to achieve a level of success no one in my community had attained. I spoke of discrimination against women and religious minorities at national conferences and international platforms. I advocated for meaningful participation of youth from developing countries in international platforms such as the Internet Governance Forum.

One of my mentors, Sultan Mehmood, taught me the importance of asking questions and the power of raising my voice. Civil society played a great role in encouraging people to ask questions. People started asking questions. This was new. This was change. The rulers and officials needed to answer and be accountable. This is a lesson I've kept with me ever since. Every time somebody says that something cannot be done, I question whether that's true. It has helped me in my projects, and probably more than any other skill I've learned.

Thanks to the Internet, my connections soon went beyond national borders. This was not just a positive. My work on women's empowerment in suburban communities meant challenging the status quo. It meant undermining the authority of traditional power structures and hierarchies. The resistance and hostility towards my work came from insecure men and women. As my work expanded, travel increased, at times even internationally. This was not considered positive by everyone, even in my family.

In 2009, my work came to an abrupt stop after I refused to marry one of my cousins. The Internet was seen as the first and immediate threat, as it had empowered me to have connections and a voice. I woke up one day and found myself under house arrest, without my modem or my laptop charger. It had not occurred to me before this incident how much the Internet had enabled me to empower myself both personally and professionally. Without connectivity, and stuck in my room, I found myself gasping for air. It was as though I had ceased to exist. I became a victim of the discriminatory practices against women I had been fighting against! For months, no one in my professional circle knew what had happened to me. Many thought I had probably gotten married. It is the norm for many women in Pakistan to stop having a career after they get married. I decided to speak up and managed by diplomacy to and through some of my forward-thinking male relatives to improve my situation. I am grateful and fortunate to have those male relatives (not that none of the women in my family are forward-thinking, but they do not have an official voice) who stood up for me and eventually got me out of the situation I was in.

I eventually could resume my work. In 2010, I was documenting a demonstration in my ancestral village near Layyah, a small city of 500,000 inhabitants, of student nurses and taking photographs of the chador-clad women sitting in the middle of the road, a rare sight in a remote city such as Layyah. Some days later, a few young ladies came to visit me.

The nurses' community was most excited about the opportunity and a big supporter of digital literacy. The reason was the digitization process that was underway at the District Hospital at the time. The nurses saw computers taking over everyday tasks, but they always had to ask the "tech guy" to come and resolve issues. They were excited about this learning opportunity at the female-only center. We solved the issue of a safe space. We engaged a female teacher, a graduate in computer science. We initially registered 25 students to train them in digital skills for office practice. The support we amassed was amazing because the business community and the local government welcomed the idea. The Social Welfare department of Punjab offered us their women's empowerment centers to host our training sessions. Soon we ran into the biggest challenge, which was beyond our control: power outages. Pakistan was going through a serious energy crisis, and we would be out of power for

6.5 Internet of Women Pakistan: Divided by Access or Skills?

18 hours on many days. This interrupted the schedule of the classes. Some of our students travelled for hours from nearby villages only to find that the center had no power. The lack of a reliable and affordable Internet connection was another major issue we faced.

The challenges we faced running the technology center are not exclusive to Pakistan. Over four billion people (60 percent of the world's population) are still not connected to the Internet, according to a 2015 report by the Web Foundation. Globally, nine out of ten people in the developing world are offline, most of them women. Women are about fifty percent less likely than men to use the Internet in poor urban communities.[2] Another recent study by the Alliance for Affordable Internet indicates that we will miss the global goal of "affordable, universal Internet access in the world's least developing countries by 2020" by 22 years, given the current trends.

The small pilot project led me to start an initiative for women around the globe to have access to technology. My initiative is called ChunriChoupaal. A chunri is a brightly colored scarf worn by women in South Asia. A choupaal is predominantly a sitting place where village elders (men) make decisions for the community members. A ChunriChoupaal Center is meant to provide a meeting place for future women leaders and change agents to come together, discuss their issues, learn and share skills. The main purpose of these centers is to provide an enabling environment for women to learn, work and meet other women. The project has received positive feedback from people around the world. We have created a non-profit organization in the Netherlands, and we are currently in the process of creating a fundraising group for the project. We have also started creating a global curriculum, which can be used and modified according to a community's local needs.

In 2014, I became a leading voice for the digital inclusion of women in disadvantaged areas of the world after becoming part of online campaigns highlighting challenges faced by community centers, such as the one I started in rural Pakistan. This gave me an opportunity to look deeply into the challenges that were preventing women from becoming part of the digital revolution. Women Weave the Web was one such campaign. This campaign collected over 600 stories from women leaders from developing and emerging countries from across the globe. Many of the challenges reported by these community leaders were similar.[3] Infrastructure, affordability, language, local culture, safety and privacy were leading challenges. This mirrored the indications from studies over the past 16 years that the digital divide was widening, even though access was available to more and more people.

This means we need something that will let connectivity and access translate to empowerment. A recent study[4] indicates that women are 25% less likely to use the Internet for job-seeking than men, and 52% less likely than men to express controversial views online in the developing world. This is attributed to the skills gap. It almost seems like a circle: there cannot be locally built apps and content unless the skills gap is addressed, and a lack of local content is a barrier for people who do not speak English.

There is a sharp contrast between the urban and rural life in Pakistan. Over 60% of Pakistan's population lives in rural areas. To access online opportunities, digital literacy is the first step for women in those areas. These areas are also more under-privileged and conservative. In cities such as Lahore, finding a female teacher is relatively less challenging than in other parts of the country. We have an extra barrier of finding trained female instructors to encourage the digital inclusion of rural women.

In Pakistan, ChunriChoupaal focuses on getting high-school girls introduced to computers and the Internet. Last year, we combined an online mentoring feature with our learning methodology to give a helping hand to local instructors who were teaching advanced skills such as coding. We partnered with the computer-science department at the University of

6.5 Internet of Women Pakistan: Divided by Access or Skills?

Victoria, Canada, and its *Women in Engineering & Computer Science* community to introduce girls to coding so that they could count themselves among the creators, and not just the consumers, of computer programs. The University appointed its first-ever digital-literacy coordinator, Veronika Irvine, to support our work in Pakistan.

ChunriChoupaal takes the view that local leaders are the key to local and regional development. An empowerment center cannot work efficiently and sustainably if there is no ownership among local people. We work with local leaders such as Rifat Arif, who runs an empowerment center in rural Gujranwala. We trained high-school girls in her center in basic digital literacy, and they are now blogging and sharing their stories with the rest of the world. Many of them are now preparing to register for an advanced course with us to learn coding and become creators.

During a speaking engagement on women's digital inclusion at the Internet Governance Forum Istanbul in 2014, an Indonesian activist asked me how we deal with aggression and opposition from men who do not support women's empowerment. The answer lies in the beliefs and actions of men such as my father, who helped me to learn and explore, and to work for women's rights. Another example is Malala Yusefzei's father, Zia, who encouraged her to blog about life under the local Taliban, which had banned female education in the SWAT valley. The answer lies in highlighting the male supporters and mentors who support girls' education and gender equality. Women need to be part of creating the technology and not just remain consumers. We need to empower grassroots-level community leaders to facilitate the inclusion of girls and women from under-represented communities.

6.6 Bridging the Digital Gap in Europe

By Iffat Gill

Iffat Gill is an international NGO leader-activist, digital strategist, and social entrepreneur working on gender equality and the economic empowerment of women through digital inclusion. She has been at the forefront of shaping the global policy debate around the digital inclusion of women. She started initiatives such as 'Work To Equality' and 'The Code To Change' to empower women to participate in the global digital economy.

You can read about the author's background and work in Pakistan on page 160.

In 2011, I moved to the Netherlands from Pakistan to join my husband. I am heavily involved in digital-literacy advocacy in Pakistan. I have spoken on it with local communities and women's groups in Amsterdam. I started telling stories about the importance of digital literacy and technology-related education for economic empowerment and gender equality. During a session in 2012 in Cybersoek, a neighborhood meeting place in Amsterdam for exchanging information on skills for jobs, including in digital technology, a young Arab woman asked, "You are interested in teaching women in the developing world about technology. Why don't you teach us first?"

I did not know how to address the question. I had initially assumed there was nothing for me to do in Europe on that front. I was soon to find out otherwise. According to a Twente University report, the digital divide is a major issue in Europe as well. Internet access for households in the European Union reached 81% in 2014 with the highest proportion recorded in the Netherlands and Luxembourg.[5] But access in the European Union is not the main cause of the digital divide, but rather digital skills and usage.

My experience mirrors this. I started speaking with more women's groups and organizations about their knowledge and the level of their digital competency. I noticed two different gaps in the digital landscape. The first was a lack of digital literacy; the second, a lack of digital know-how at the organizational level. The demographic most affected by this is predominantly generation X, baby boomers, or people who have immigrated from different countries—that is, precisely those who are often the least integrated in the Netherlands.

I began working with community leaders who were involved in the empowerment of local women's communities. Both the leaders and their target audiences often found themselves facing technological issues because of a lack of proper training. Most of these women relied on younger friends, children or grandchildren to explain how certain software or programs worked.

Digital literacy includes fluency in digital systems, the ability to use a range of tech tools to accomplish work-related tasks, and the flexibility to adapt as technology changes. Digital competence has been acknowledged as one of the eight key competences for lifelong learning by the European Union. Digital competence can be broadly defined as the confident, critical and creative use of ICT to achieve goals related to work, employability, learning, leisure, inclusion and/or participation in society.

This demographic was restricted by such simple tasks as setting up a Google or a LinkedIn account. People may have the equipment, but not the knowledge to use it beyond the basic word-processing tools. I started volunteering to help out women's rights groups and organizations. Community organizations were interested in recruiting me because of my digital skills; setting up e-mail accounts, social-media accounts, setting up a WordPress website, and retrieving documents from their computers.

I upgraded my own digital skills and set up my consultancy business to help non-profits achieve their social-fundraising goals and have a strong online presence that is aligned with their organizational needs. I started with small organizations such as African Women Perspective (AWP), an organization run by Dutch women of Ethiopian origins that advocates for the inclusion of the African diaspora in Amsterdam. Organizations such as the AWP had no online web presence until 2012 because they ran on a small budget and did not know the right people to help them out with it.

Setting up a digital strategy for organizations became the focus of my consulting business in the Netherlands. I helped organizations launch their social-media departments to maximize their impact. The bulk of my work was tweaking their media strategies to achieve their fundraising goals, or help with community-building and engagement. This demographic uses modern ICT tools in their work on a regular basis. But they never learnt about the repercussions of using these tools. Examples of the skills gap came not just from these conversations, but also from how the organizations communicate. The lack of training on "netiquette" (Internet etiquette) is pervasive. For instance, I came across many small and medium organizations that do not use "BCC" while sending a message to multiple people, and inadvertently end up violating other people's privacy. When I inquired about their privacy policies, one organization sent me a long generic e-mail about how valuable their members' privacy is to them, even as they were in fact violating that privacy. And that was not an isolated incident; many organizations and networks make the same inadvertent mistake. This demonstrates that the digital competency of a lot of organizations needs work.

Unfortunately, for many of these organizations, digital skills and digital competency and related training are not high on their priority list. Indeed, even if it *is* seen as a priority, there is little knowledge about how to acquire these skills and competencies.

A 2012 report by the European Literacy Policy Network confirms that, although schools and/or public libraries have invested in digital media, many users do not have the knowledge, the skills, or the access to training to use them effectively.[6]

Women in the Workplace

In the course of my consultancy work, I met with many women with whom organizations such as AWP worked. Their monthly meetings were frequented by community members; Dutch women both native and from countries such as Afghanistan, Iran, Egypt, Ethiopia, Pakistan, India, Ghana, Suriname, Turkey and Morocco. These women showed great interest in digital tools for improving lives, and had a keen interest in learning about smartphones and

6.6 Bridging the Digital Gap in Europe

tablets, for instance. They wanted to be up to date on modern communication tools, so they would not have to rely on others to show them how to check their e-mail or do their online banking. Women want to break free from their technological dependence and take control of their lives.

I kept getting requests to bring digital-literacy training to women. My network of women's groups and organizations kept telling me that there are many talented women who want to look for an "in" to the job market, and that the bulk of the jobs available were in the technology sector. The problem was the skills gap.

I realized that this was not something that I could manage as an NGO, so I started working to include other stakeholders, especially from the private sector. In the winter of 2014, after I had returned from a meeting of the UN's Internet Governance Forum, I issued a call to women working in the technology sector to join forces to bridge the digital-skills gap. I founded the Amsterdam Women in Technology meet-up group. The meet-up is open to everyone who is interested in becoming part of the gender-diversity debate and who is willing to take action. Members are from all professional levels in technology companies.

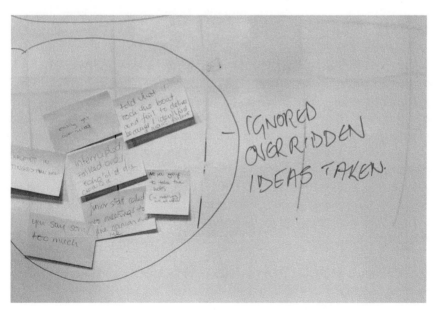

Figure 6.1 "Cracking the Code" group activity.

(photo: Eliane Kemble)

The first key supporter to respond to my call was Mine Ogura, a senior technologist at Marktplaats (e-Bay Netherlands) who had played a key role in hiring more women programmers in her own company. I started working with her and built a team to find solutions to address the problem, which I had repeatedly encountered. We launched a series of meet-ups to involve technologists from the industry in our project and to share knowledge with others. The meet-ups took a different and unexpected turn!

In our first meet-up, in March 2015, one member told her story. She and her husband had been hired together by an online travel company. They were both equally qualified programmers. The hiring manager said to the husband that he would give him a higher salary than his wife because *a man is supposed to earn more than a woman*. Yes, that was 2015—in the Netherlands, a country that actually has a policy on gender equality and a long tradition in emancipation policies to promote equal rights, equal opportunities, equal liberties and shared responsibilities for women and men.

This indicated how important the role of culture and mindsets are. The Netherlands Institute for Human Rights reports that stereotypes (implicit and explicit bias) are an important cause of employer's discrimination in the recruitment and selection of new staff.[7]

In an effort to record some testimonies from women working in the technology sector and getting some data to find a good starting point to work from, I started a campaign called "Work to Equality." This campaign was launched to share stories and thoughts from women who are working in STEM-related fields to raise awareness and help create better working conditions through understanding. The idea behind this was that, if more women are happy in their jobs, they will be more likely to attract new talent and encourage girls and women to join and follow their path to economic empowerment. This was crucial for the work we were doing to train new women who wanted to explore opportunities in the technology sector. The data we collected was both from our "Cracking the Code" session during our Women in Tech meet-ups in Amsterdam, and from our online questionnaire for joining Women in Tech. Anywhere from 20 to 50 women and men from the industry generally attend our meet-ups. We have over 220 members in our meet-up group at the time of writing.

Here are some quotes that give an indication of what women in the industry have said during our meet-ups, and that illustrate that we still have a long way to go.

> "Women always get this question: So, do you have kids? Males don't get asked this question much."

> "I get interrupted 10 times more than my male counterparts in company meetings."
>
> "Sexist jokes by male colleagues make me feel uncomfortable and confused. I do not know whether to laugh or feel offended."
>
> "I am considered disruptive if I express my opinion about something I do not agree with."

In our analysis of the responses from over 200 women, we found that they attributed the challenges to conscious and unconscious bias, which stems from the current education system, people's upbringing, and cultural norms.

Here are a few responses we got from the question on challenges for women in the technology sector.

> "Education systems make technology seem a more 'manly' profession that deters women from entering STEM roles."
>
> "There is a lack of opportunities for beginners to train on the job, because companies prefer not to invest in learning (though that seems like a generic issue as opposed to only a women's issue)."
>
> "Lack of women in this male-dominated [technology] field is caused by the perception of women in media and society."
>
> "Lack of mentors and role models."
>
> "Work cultures that don't support family life."
>
> "To be listened to on technical solutions as though I were a male developer, without the quizzical look that says, 'Are-you-sure?'"
>
> "The biggest challenge is for the men, who need to change their attitude to be more welcoming to women."

If the working environment is unwelcoming, attracting new talent will be a huge challenge. Our meet-ups in Amsterdam are attended by many curious men and women who are interested and invested in creating change around them and in playing an active part in getting more diversity into technology fields.

The Gender Skills Gap and Economic Empowerment

ICT jobs in Europe go unfilled because of difficulties finding qualified candidates. We need to focus on what can be done today and on how to solve the lack of diversity and inclusion in the workforce.

In addition to women in technology, our meet-ups also got the interest of people who wanted to join the technology field and were looking for a way in.

Many women wanted to join the technology sector, but did not know where to start. These people came from different educational backgrounds. They were unaware of the opportunities available in the ICT sector, and had no mentor or support group to find answers to their questions. They were interested in upgrading their skills for IT jobs but were unsure how to go about it. There are countless educational resources available online. But if you have no previous knowledge, how do you know about different coding languages and what kind of jobs you can get access to if you do manage to build your skills in that language?

Many women join one-day coding bootcamps such as the one offered by groups like Django Girls. While they may have positive experiences at the bootcamp itself, the road ahead from there on remains unclear to many attendees. While a small percentage might keep on learning by themselves, many feel lost and even more overwhelmed without anyone to mentor or guide them. One trend I noticed, after graduating from one of these bootcamps myself, is that recruiters start sending messages to the organizers about job openings. Observing and taking part in these bootcamps made it apparent that, in order to get job-ready people, a more extensive program is needed that can tackle the skills gap and the lack of diversity in the industry. With all the trees that the Internet offers, it is easy to lose sight of the forest. What's needed is a guide.

After extensive talks and interviews, with groups such as Women in between Jobs, we launched the Code to Change program. The program kicked off with an informational evening about different opportunities and fields that exist in the ICT sector at the entry level. The evening gave people clear insights into how learning new digital skills can complement existing skills and learning interests.

Age Group

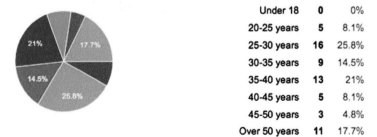

Figure 6.2 Age diversity of the Code to Change 2015 applicants.

We decided to put Women in Tech and Women in between Jobs in touch with one another. The Code to Change program connected women who were looking for opportunities to join the IT industry with women who were already in the ICT sector and were willing to teach others.

With a small team of three people and a handful of volunteers, we put out a call for applications in summer 2015. We received 62 applications, and their geographic diversity was impressive. In addition to women in the Netherlands, we received applications from women in Argentina, the United Kingdom, Pakistan, Indonesia, Serbia, Nepal, Kenya, Germany, Belgium, Zimbabwe and Uganda. (Some applicants did not specify where they were based.)

Applicants' ages ranged from 20 to over 50.

Going through the applications, we noticed the following reasons for joining the Code to Change program from applicants:

1. Economic empowerment; build digital skills to explore careers in the technology sector. "My goal, to excel in what I learn here to get employed." "My goal is to learn new skills in order to re-enter the job market."
2. Become part of creating technology rather than just continuing to be a consumer: "Working 'only' on the commercial side of product development I watched with envy how my (mostly male) colleagues build stuff. I want in!"
3. Career switch from other fields within the technology sector: "I believed that hard-core programming was something only guys did and that I probably would not become a good enough programmer. However, this idea changed over time. I have a law degree and work as an administrative manager surrounded by programmers. I feel that I can add value to the company by learning how to program."
4. Replicate the program: "My expectations are to get to know how to generate a friendly space when it comes to learn how to code (and also learn how to code myself). And use that knowledge to replicate the experience here in Argentina for workshops for women."

With the support of women technologists and corporate and non-profit partners, 30 participants, belonging to the first three interest groups above, were selected for the first phase. We decided against applicants from interest group four in the pilot phase: many of the participants who applied from this group were from the non-profit sector, and this meant they had limited resources to pay for the travel and accommodations if they were to fly in from other continents. Unfortunately, we had no means to support that.

192 *Reshaping Traditional Ways of Learning*

The participants we selected were from the Netherlands, Belgium and the United Kingdom, of diverse ethnicities and professional backgrounds. The participants included a PhD student, a marketing professional, an accountant, lawyers, sales and communications assistants, a financial analyst, a waitress, and a nanny. They came from such countries as Suriname, Pakistan, India, Egypt, Morocco, the Netherlands (native Dutch and of Indonesian descent), Croatia, the United Kingdom, Bulgaria, Belgium and Brazil.

The Code to Change kick-off event comprised a two-day digital-skills bootcamp and a one-day conference. The main aim was to provide a safe learning environment, and the bulk of the activities were organized around building confidence with technology, team, and community, and building a community of supportive women who wanted to learn, grow, connect and code. The event was interspersed with lightning talks and lectures on both professional development and various technical fields.

We managed to get our participants excited about technology. The following are some of the reactions we got from them:

"I've got a job! It's quite incredible. I've landed a really nice (tech-focused) technical-writer role. This was entirely thanks to the Code to Change."

Figure 6.3 One of our mentors, Alexandra Vargas Ortega, with the Code to Change 2015 mentee Kauser Khan, Amsterdam, the Netherlands.

(photo: Arno Meulenkamp/the Code to Change)

"Everyone is afraid of losing their job. The Code to Change workshop helped me see that I'm not on my own: I have a group of people who will help me on my new journey."

"I felt part of a small community of very enthusiastic women, and I felt taken seriously in my ambitions to move towards the tech industry for the first time. That made me feel empowered to actually start and take the first steps towards this goal. The event was a great mix of coding, speeches, talks and other get-togethers."

The lesson we learnt as an organization was that at one end there are so many vacancies that need to be filled in the technology industry. The shortage of skilled professionals is expected to run through at least 2020, not to mention the updating of the existing skills of people already in the sector. On top of that, it has become apparent that the formal education system in many countries is not producing graduates that are job-ready coders. Programs such as the Code to Change are practical training grounds that can help fill this gap.

There are a great number of talented women who have this hunger for knowledge and want to become part of the digital economy. And yet, not enough investment is going into the existing solutions or programs such as the Code to Change, from the public or the private sector.

A recent report on coding bootcamps by the International Telecommunication Union (ITU) states that they are a "space to watch and to explore" as they work to bridge the digital-skills gap in part through the inclusion of youth and women.[8] The report focused on programs such as the Code to Change, which are three- to six-month intensive, hands-on, in-person training programs to learn coding. The report concludes that it is imperative that policy-makers and other stakeholders explore every avenue—including emergent opportunities such as the bootcamp model—to mitigate this condition and improve the earning prospects of youth worldwide.

We are moving in the right direction, but we still have a long way to go. The time is now to work towards equality on many fronts, from the subconscious bias to push girls towards certain subjects to the culture in ICT companies that makes women that do choose careers in ICT leave the field in much larger percentages than men (leaky pipeline). Women working in the technology field know what challenges they face, and may have an idea about what the potential solutions might be, but they may not have all the answers. While it is important to listen to them, it is unfair to burden them alone with solving the conscious and unconscious biases ingrained in our society.

Goodwill alone is not going to solve the problem. We need more investment and a greater willingness from the industry to make newcomers feel welcome in the industry, and that includes opportunities for on-the-job training, apprenticeships and mentoring programs. Having more women in ICT not only helps new women getting an exciting new career, it also will help retain the women that are already there by making the working environment more inclusive. Most importantly, it will help the ICT sector as a whole, as diversity makes any team greater than the sum of its parts.

What I have learned throughout my career is that we all need support. We all need mentors to guide us through the barriers that might stop us from moving forward on our learning journey. Learning a new skill can be scary, and self-doubt can make it a crippling experience. We are amazed at the amount of support the tech community has shown us because of their desire to give back, to pay it forward. We need to leverage this drive to "give back" to bring positive and lasting change and to support others in achieving their goals.

And unwelcoming work environment, biased hiring practices, and wage disparity stop the existing female workforce from growing in technology careers. The growing number of job vacancies in the ICT sector indicates that we need to fill these positions now. Reintegrating women into the workforce by plugging the skills gap can help tackle unemployment and the employee shortage in the industry. Therefore, we need to level the playing field. More action is needed from the technology sector to ensure women's empowerment and greater gender equality. It is also important to mention the crucial role to be played by women's organizations, networks and female-led start-ups that are already working to address the issue of the digital gender divide and the digital-skills gap. They are already working in the field, experimenting with the solutions. Collaborating with them can only contribute to accelerating change.

The Netherlands is performing well, according to the European Gender Equality Index. However, the report on gender-equality policy concluded that the position of men and women in the Netherlands is not equal and that there are still several issues to tackle.[9]

And this is not about feminism, affirmative action or activism. It is about improving productivity (mixed-gender teams perform better than single-gender teams). It is about economic growth. It is about getting Europe (and by extension the world) and everyone in it ready for the future that is already upon us.

References

[1] *"Malta – Literacy rate". Indexmundi.com:* http://www.indexmundi.com/facts/malta/literacy-rate

[2] Global Report-October 2015: Women Rights Online by Web Foundation: http://webfoundation.org/wp-content/uploads/2015/10/womens-rights-online21102015.pdf

[3] http://www.worldpulse.com/en/get-involved/take-action/250

[4] http://webfoundation.org/wp-content/uploads/2015/10/WROinfographic.png

[5] Eurostat 2015, Information Society Statistics – households and individuals: http://ec.europa.eu/eurostat/statistics-explained/index.php/Information_society_statistics_-_households_and_individuals

[6] http://www.eli-net.eu/fileadmin/ELINET/Redaktion/user_upload/LITERACY_SUMMARY_EN.pdf

[7] https://mensenrechten.nl/sites/default/files/2013-08-23_Strategic_Plan-English.pdf

[8] Digital Inclusion, Coding Bootcamps: A Strategy for Youth Unemployment: https://www.itu.int/en/ITU-D/Digital-Inclusion/Youth-and-Children/Documents/CodingBootcamps_E.pdf

[9] http://www.europarl.europa.eu/RegData/etudes/IDAN/2015/519227/IPOL_IDA(2015)519227_EN.pdf

7
Educating Women in Post-Conflict Zones

7.1 Developing a Technology Community for Women in Afghanistan

By Fereshteh Forough

Fereshteh Forough is the Founder and President of Code to Inspire, the first coding school for girls in Afghanistan. Fereshteh was born as an Afghan refugee in Iran. She has a master's degree from Technical University of Berlin in Germany. Her passion is to empower young women from Afghanistan by improving their technical literacy.

In November 2015, we opened Code to Inspire, the first coding school for women in Afghanistan. I have an advisory board from Google, IBM, ConsenSys, and others in the tech sector. In Afghanistan, we currently have 53 students—35 of whom are in high school, while the rest are enrolled in university computer-science programs. I really wanted to establish Code to Inspire because the technology sector is just forming in Afghanistan.

In Afghanistan, there are many struggles for female computer-science majors. Imagine graduating with a computer science degree and trying to find a job. There are certain barriers that really make it difficult. First of all, a lot of families are conservative when it comes to where a woman may work. The majority prefer that their daughter become a teacher since it's a well-respected job in society and they'll come into contact only with other women. Most women can't find jobs in computer science. The only thing they can do is teach and it probably won't even be in computer science, it'll be math, English, or history. If the school does need a computer teacher, it won't be for a very technical curriculum, but to teach how to use program packages such as Microsoft Office.

If you want to find a job outside your hometown, there are cultural issues to consider, especially for women. Most families will not let their daughters travel alone by themselves unless they have a male companion. Furthermore, most families wouldn't be able to afford a plane ticket. It's also not part of our culture to go to another city and stay with someone who is a roommate unless they're also a trusted family member.

The other issue has to do with entrepreneurship. If you want to be an entrepreneur and create your own company, when you reach out to a potential customers and tell them that you can design a website, they often say, "You're a woman, do you think that you're able to do that?"

All these things make it very difficult for young women who graduate from computer-science programs to pursue careers in what they studied. That's what inspired me to create Code to Inspire as a safe space for women to come and build their skills. It's free of charge, and we have created a specialized curriculum. For example, the students in high school are learning basic web design, HTML, and CSS. Our mission is that they should graduate from our coding school and select computer science as their major in university.

Currently, in Afghanistan, there is not a lot of awareness about technology and little encouragement for women to go into any technical field. In the second year of our program, we plan to offer more-advanced website design. Our goal is for graduates of our two-year program to become professional website designers and take jobs both within their communities and virtually. Our students are between the ages of 14 and 18.

We are also developing a curriculum to teach students how to build mobile applications. Today 80 percent of people in Afghanistan have access to a mobile phone either directly or through a family member. The majority of young people have smartphones, but Internet penetration is only seven percent. On mobile, it's a good market and connectivity is steadily improving throughout the country.

To develop mobile applications, students learn how to develop an idea from scratch and work as a team while learning different technical aspects and technologies. But then there are the cultural aspects, and a student can identify a problem in society and think, "By developing this app, we can solve that problem or make it easier for people, and develop it in our local languages, Dari and Pashto." So once people download the app, especially men who don't have trust in a woman's ability and use that app, they're going to say, "Wow, women did that." So it's an empowerment in itself that sends a message to people.

We start with smart phones and mainly Android because not a lot of people have an iOS phone in Afghanistan. I'm trying to find corporations, companies, or organizations that can offer internships to our students or sponsor them for

7.1 Developing a Technology Community for Women in Afghanistan 199

a period of time. After the internship, I hope they can hire them even part-time to do very simple jobs, even maintaining or designing websites. Previously when Afghanistan was flooded with donor aid, many educational programs were created, but graduates could not find jobs. I don't want this to happen with my program. I want the students to be able to sustain themselves after they graduate.

We had 120 applicants, and from among them we picked the best ones who had good English skills and basic computer knowledge. We didn't want to teach them how to turn on a computer or how to create a folder. Every year, we aim to have 50 new girls registered, because our space is small. My long-term plan for the next three years is that we open new coding schools in Mazar, Kabul, and any other cities in Afghanistan where there's demand. I want to create a strong community of women in tech so that Code to Inspire hubs can support each other and help each other. We can have annual conferences, workshops, and the young women who graduate can be mentors to those who follow them in the program.

That's the plan for the next 3 to 5 years. And hopefully we will succeed. We did an online fundraiser last summer through Indiegogo. Our goal was to raise $20,000, and we raised $22,000. We got 20 laptops as an in-kind donation from Overstock.com. We also got a donation from the Malala Fund. I met their CEO, Meighan Stone, at the Clinton Global Initiative conference. Since two of my board members are from Google, they helped us get on the not-for-profit Google list that provided matching donations.

I really want to turn our not-for-profit mainly into a social enterprise so it becomes more like a business incubator. Once students find a job through us, they can contribute a small percentage of that to the foundation. One of my board members and the co-founder of Code to Inspire is John Lilic from ConsenSys, a company focused on blockchain development. I think about identification in Afghanistan because not a lot people have an ID card or passport. A lot of people don't even know where they were born. They're like displaced people. This makes many people unable to open a bank account or use PayPal.

With blockchain technology, the key factor is the Internet. If you have the Internet, you can create a wallet for yourself with your username and password—that's it. You don't need all the additional documentation or need to have a middle or a third party involved. It's very good for a lot of people sending or receiving small amounts in the range of even $1, $2 or $5. The banks and even some mobile payments charge high fees for transactions, which end up being a lot of money for a family in Afghanistan.

It's definitely a better option for a lot of people who live overseas as well and want to send money to their families. Sending money through Western Union or a bank comes with high costs. The financial aspect is one thing about Bitcoin, but blockchain technology can be used in any kind of sector in multiple ways. I think with the programming side of blockchain, as our students become more skilled, I'd like them to work on applications that could help improve and solve problems in our society.

The telecommunications sector in Afghanistan has really improved with 3G Internet expanding. The NATO-funded Silk Road for Fiber Optic Project is going to be finished so the Internet will become faster and cheaper. It's a circle going from Herat, Kabul, and Mazar and connecting Afghanistan's main cities. In Afghanistan the information technology sector really progresses most in comparison to any other sector in the country.

Although Afghanistan faces many challenges, I'm still hopeful. A lot of people in my age range were refugees during the USSR invasion, and lived in Iran, Pakistan, and other countries. Those who stayed in Afghanistan experienced being with Taliban and definitely don't want them to come back. They want a different future.

My advice to young women is to engage their local community and families. When you make a connection and build bonds of trust with families, they'll really support you. It's important for any program to win the trust of the community it's being run in. Once the community trusts you, they may be an advocate for you. Like the parents can go and talk about how they sent their daughter to a coding school and how she's learning amazing skills. That way, other parents will want to send their daughters, too.

7.2 Overcoming the Wounds of Genocide in Rwanda through Education and Technology

By Alphonsine Imaniraguha, with supporting research by Cindy Cooley

Alphonsine is Founder and CEO of Rising Above the Storms and a Network Consulting Engineer at Cisco Systems. Cindy Cooley works for Cisco Systems and is Co-Chair of TechWomen.

In the post-genocide constitution of Rwanda, 30% of the seats in the lower house of parliament are reserved for women. Only two countries in the world have a higher percentage of women serving in parliament than men. Rwanda tops this list with approximately 64% female representation. Through this leadership there have been policies to protect women against violence and create opportunities for gender equality. Rwanda scored .7854 out of a possible equality score of one, which ranks number seven out of 142 countries in the Global Gender Gap Index 2014 conducted by the World Economic Forum.

The African continent is the second largest in the world in terms of land mass. Africa is made up of 57 countries. The African continent has huge untapped potential of 900 million consumers.[1] Africa must produce, not just consume technology.[2]

Rwanda is seeking to be known as a knowledge capital for STEM education and skilled workers in ICT. One of the goals of this small, landlocked country with few natural resources is to develop a skilled ICT workforce that can also be considered as a hub of exporting ICT human capital to other parts of Africa.

I was born in the suburbs of Kigali, the capital of Rwanda. Rwanda means, "getting bigger or enlarging." When I was growing up, everyone around me thought that Rwanda was the biggest country on the entire planet, when in reality, it is about the size of the tiny United States State of Vermont.

I was born in a big family, the second oldest daughter of six children—two boys and four girls. My first name, Alphonsine is French. It means a "noble warrior." My last name, Imaniraguha, is a Kinyarwanda (my native language) name. It means, "God gives you." Most Rwandans to this day don't always carry their family last name. Imaniraguha is my unique last name; my siblings have different last names.

Although my parents had no more than a basic education, they were aware of the value of an education, and while it wasn't very common to send children to pre-school, my parents did, for all their children, except me. My parents did not send me because I had already learned everything they taught in pre-school and much more. My dad had taught me about all African countries, their capital cities and their presidents. I knew most of them by the time I was five. He also taught me the English words he knew and how to count, although he didn't even have a high-school education. When I was six, my parents sent me to primary school.

From first to sixth grade, I was always at the top of my class. I loved school and never wanted to miss a single day. I vividly remember one time I had a motorcycle accident in the final semester of first grade, which almost cost me my whole right foot. My greatest grief was not my pain or the fact that I had to be carried like a child at six, since I couldn't walk. No—I was heartbroken because I would have to miss not just a few weeks of school, but a whole semester and possibly not move to the next grade. I couldn't handle that.

I sent someone to ask my teacher if she could send the final exams and allow me to do them at home. She agreed and when I received the exams, I sat in my room by myself and did everything, alone. When I was graded, I got everything correct and my cumulative grade was 99%, although I am pretty sure my teachers always took one point off somewhere so that I don't always get 100% every year.

In the third grade, my teacher politely asked me to substitute for her for three days when she took a vacation. I enjoyed helping my classmates so I agreed without hesitation. It was not long until our neighbors came to my dad and said, "Guess what, instead of learning as a student, your daughter is actually the one who teaches other students." My father didn't get mad. Instead he replied, I learned later, "If she has something to offer to others, I don't see why she should not help."

My dad always had confidence in me. To this day in Rwanda, at the end of the sixth grade, you have to take a state exam to be able to go on to high school. Prior to the exam, you have to fill out forms to indicate what schools and major(s) you would like to take, should you pass the exam. Based on the score and subjects you did well in, the Ministry of Education places you with a school and assigns a major.

As a child, I thought it was cool to be a nurse or a teacher, because those are the people I saw often. However, when I got my state exam results back, the Ministry of Education itself decided that I didn't belong to nursing or teaching. My major for the next six years of high school was going to be mathematics and physics. Moreover, I placed into the top high school in the whole country. Little did I know then that this was the beginning of my journey as a woman in technology. My parents were overwhelmed with pride.

Late in March 1994, around my 13th birthday, nothing could ever have prepared me for what was about to happen. Right after Easter, on April 6, the plane carrying the president of Rwanda was shot down as it was landing at Kigali International Airport. Within seconds, the genocide against the minority Tutsi group began. Over the next three months, around a million Tutsis were mercilessly slaughtered. This staggering number includes my parents and two of my siblings. A stranger saved me. My three younger siblings, who were eight, six, and three at the time also survived.

When the genocide ended, life returned to being as normal as it could be. I eventually returned to school. With a drastic turn of events of this magnitude, and no counseling centers or qualified therapists, every single person had to find their own way of coping with this unthinkable turn of events. Many orphan girls my age got married or simply became prostitutes. Many ended up with diseases, and their lives were pretty much over. I, on the other hand, joined a religious group, only to discover later that it was a cult. Without going into details, the group discouraged us from studying or taking medications when we were sick. We were only encouraged to pray. I was there for three years.

While I had always excelled at every class subject, my grades started to decline in high school. However, slowly, I eventually discovered the lies of the cult and how they were using the loss and grief we endured to lure us to believe what they preached. When I eventually left the cult, it was at the end of high school in Rwanda when you have to take another state exam that would qualify you for college. The exam covers everything learned in the final three years of high school, which so happened to be the time I missed most of my classes.

This exam would decide the fate of a future I already lost hope for. Since there was no way around it, I gathered all the books and materials to teach myself math, physics, biology, chemistry and the courses covered while I was busy searching for God. I had exactly 21 days to learn three years' worth of course work and had no idea where to begin. For the next three weeks, surprisingly, I was able to sleep for only one to two hours a day, and I didn't feel tired at all.

Three weeks passed by, and I took about six different exams. When results came in a few months later, my results were so high, I received a full scholarship to study Electronics Engineering at Kigali Institute of Science and Technology, in Rwanda. This very moment of hope shaped who I am today with my personal faith in God and life. In college, I was determined to study hard and didn't want to be affiliated with any religious groups.

However, a post-traumatic stress disorder that I had perhaps ignored caught up with me. I was diagnosed with a stomach disease and an ulcer. In college, no one believed that I would ever graduate because I was hospitalized at least every week. Every medication that was available in Rwanda that could help was tested on me. When I felt better, I would simply return to my classes to take exams after spending days in the hospital bed. Miraculously, I would do so well that my classmates, who were all male, thought the teachers were giving me high grades because they favored me. In my senior year, the disease finally was healed.

When college graduation came, I graduated with honors and was offered a full-time job immediately. Three months into the job, the Rwandan Ministry of Education received two scholarships for master's programs from the Rochester Institute of Technology in the United States. I was very thankful to be one of the recipients, and a few months later, I packed my bags and moved to Rochester, New York.

When I arrived in the United States, the English conversation was a really big problem for me at first. Although my undergraduate degree was taught in English, we didn't speak the language outside the classroom or to carry a normal conversation. Plus, I didn't understand the American accent. It was a challenge. However, as always, when I am determined, nothing can stop me from reaching my goals. I took English for 10 weeks and started graduate courses after that. I majored in Telecommunications Engineering. To my surprise, and also my sponsors', I completed the first semester with straight As and a GPA of 4.0.

During the graduate program, I was exposed to the Cisco lab on campus; the tasks were learning how Cisco infrastructure works, how to configure routers and switches, and get an IP phone to connect to the network and make a conversation. That itself developed my new dream: Cisco was going to be my destination for my professional career. The only problem was, how would I possibly get Cisco to hire me? English is neither my first nor my second language—it's my third. Moreover, I had no experience in the field, or strong connections. And to cap it all, the economy crashed in 2008 around my graduation, so that made the process hard. Many companies froze new positions.

When I was first selected for the interview process, Cisco, like many companies at the time, was having a hiring freeze. That didn't stop me. I tried again later. I navigated the challenges and eventually was offered the job I always wanted in the summer of 2010, a Network Consulting Engineer. I was ecstatic. My thesis advisor at RIT later told me that I am a living proof that someone can pursue something great and achieve it.

Today I see the world differently; I believe that anyone can become whoever they want to be. While my parents always believed in me as a little girl and were confident that I would do well in whatever I chose, when they were killed, that support and encouragement I received disappeared. Throughout high school and college in Rwanda, no one had ever told me that they were proud of me or encouraged me to do well. In fact, it was the opposite: the high-school cult leader and one of my aunts harshly told me, on separate occasions, that I am good for nothing and that I would never get anywhere. I never knew what I was capable of.

Giving back to Rwanda through Technology

Education brought me to the United States, landed me a great job, and gave confidence to an orphaned girl. I feel grateful and compelled to start my nonprofit. Our vision is to provide a safe haven for at-risk youth and street children to dream, be told that they have the ability to become whoever they want to be, and hear that they can make someone proud. Our first project is to be launched in Rwanda in 2017: a learning center as a hub for education and empowerment. I want them to know that, just because they may feel like outcasts, this doesn't mean that it's the end. I am living proof of that.

I am thankful to have been able to raise and fully financially support my three siblings. Two of them have completed their master's degrees; my youngest sister is done with college and waiting to start the graduate program. In my household, education comes first. It's a rule, not an option!

7.3 How Technology Is Connecting Communities of Women in the DRC

By Sarah Thontwa, Senior Research Assistant at the International Food Policy Research Institute (IFPRI) and Co-Founder of Femmes Politiques (FEMPO)

The Democratic Republic of the Congo (DRC) has recently come out of years of civil conflict that claimed over six million lives, thus making it the most deadly conflict since World War II. The DRC held its first two democratic elections in 2006 and 2011, thus indicating the country's decisiveness in moving towards reconciliation and nation-building.

I was born and raised in the DRC. My early upbringing alerted me to gender inequality. Instead of burdening me with household chores, my parents insisted that I go to school, complete my homework every day, and read books. It soon became clear that my normal was not so normal for many of my classmates who were girls. After school, they were expected to fetch water, cook, clean, and service their households, while their brothers simply went to play. By the time I entered high school, my mother was often ridiculed in the community for raising her daughter to read books and for gambling with my future. It was not important for girls to be smart, they said—what mattered was for them to make good wives.

I was privileged to have parents who believed in education and who gave all they had to help me chart a new path. As I grew older, I wanted to support girls' dreams and ambitions as a way to address the challenges that largely prevented my classmates from excelling in school, completing their education, and finding a place of their own in society.

As a student in college, I started Littlethings, an initiative recruiting smart girls in the communities where I grew up, providing them with scholarships, leadership, and support for their personal-development so they could follow their dreams. This initiative later grew to become a leadership incubator that invested in the personal, professional, and educational development of young women in the DRC. Over the years, I have continued to invest in the development of women's capacity and in the widening of opportunities available to them through my work as advocate, political trainer of women in politics, and fundraiser to support high-impact initiatives by women in the community.

Being a fervent advocate for women's progress and advancement, I have continued to pay close attention to the place and space occupied by, and attributed to, women during the course of this transition. The DRC ranked 149 on the 2015 Gender Inequality Index report. Women's and girls' progress is challenged by factors such as a discriminatory family code, restricted civil liberties, cultural bias, and restricted access to finance and resources. Women occupy only 8.2% of seats in the parliament and participate at a lower level within the government than their male counterparts.

However, despite strident barriers to women's development, Congolese women are resilient. They are actively involved in the process of transforming their communities, and are determined to bring change to their society. As DRC shifts its focus from being at war to being in a post-conflict phase where it seeks stability, prosperity, and economic growth, women are navigating old and new social political ground by using some of the best technological tools of our time to engage in advocacy and to address their challenges.

The Internet penetration rate in the DRC currently stands at just above 3%[3] with a mobile-phone penetration rate estimated at 23%[4], with the majority of users living in urban areas. In the DRC, as in most other low-income countries, mobile phones are cheap and easy to use. They operate under a built-in price-per-second model and are adapted to local market needs. Users need only basic numeracy and literacy skills. Largely influenced by Chinese manufactured products made for African markets, these cheap and reliable cell phones provide multiple sets of interactive features via mobile apps, text messaging, voice calling, Internet browsing, and social-media access. In a country of over 70 million inhabitants, the ability to access the Internet and social media via cheap phones at a lower cost has catapulted the DRC to the top 10 African countries with the fastest growth of Facebook users.

It is largely men who are represented in the Internet and social-media space because their economic status is higher than that of women. However, Congolese women are increasingly finding ways to engage in this space alongside their male counterparts. Access to communication tools and technology has become an important medium for Congolese women to tell their stories and contribute to the making of a new era.

In 2015, I was privileged to meet a diverse pool of Congolese women who in their own right were influential within various political parties. This cohort of women at the time were taking part in ensuring election preparedness and in the political training provided by the Women's Political Academy of the National Democratic Institute (NDI), where I served as political trainer and consultant. Committed to building equal participation by women in politics and government, the academy had the goal to equip female candidates running in the local elections with the fundamental tools, skillset, and political know-how to run effectively and succeed in politics.

Under the leadership of Eve Thompson, former Country Director of NDI in the DRC, candidates were selected across political parties following a rigorous selection process. Irrespective of political affiliations and platforms, the marginalization of women and the condescending attitude towards their voices are recurrent themes. For Congolese women who are building change, this has become a unifying factor that prompts women to build blocks of solidarity as they struggle to transform communities and shift national policies.

The academy introduced these women to tech tools and trained them on how to blog as a way to insert themselves in the national, social and political dialogues. The value added by, and the benefit of, Internet training and blogging for women in the DRC can be seen at three levels.

At the first level, women in the DRC who struggle to make their voices heard and to highlight their platforms are finding additional options by making the public and their leaders pay attention through the Internet. Unlike the reality of the physical environment in which they live, access to the Internet has given them a platform where their experiences and voices do not need to be pre-validated by men who maintain heavy social control and political influence within their political parties.

On a second level, Congolese female politicians, civil-society leaders, and bloggers are able to directly channel and voice their constituents' needs to decision-makers. Policy-makers, most of whom are men, are located primarily in urban areas and are often disconnected from the reality of the most marginalized. The feminization of poverty has translated to women being among the poorest of the poor. Congolese women trail behind Congolese men in almost all indicators of development. Women and girls hold lower status in health, education, political representation, and in the economy.

Because of discrimination, cultural constraints, and perhaps a relative lack of adequate funding, these negative repercussions on development are not always clearly visible to decision-makers, and whenever voiced by women, such matters are not given the attention required to implement corrective policies. Blogging has given Congolese women a platform for continual advocacy as they are consistently able to put development issues affecting their communities on the map, thus forcing policy-makers to pay attention and keeping them accountable. Lastly, blogging has allowed Congolese women to be seen at the frontline of the effort to bring about change, as opposed to being seen as shadows. As the public interacts with these women online and offline, they are able to meet women who bravely articulate their platforms and whose ambitions and actions are changing their communities.

There are currently 25 women working on the DRC Molongi Blog initiative funded by USAID. They are blogging on issues ranging from the electoral process to gender equity, and from economic access to education. Anne Marie Makambo was named "best blogger of the year" in 2015. She discovered blogging and the Internet while taking part in a training program conducted by the NDI. She started blogging almost immediately, promoting her political platform where she speaks out on societal issues pertaining to women and youth. She has become a vital voice addressing the opportunities and challenges shaping women's experiences while navigating the political terrain in the DRC. She highlights realities from the communities she serves, and promotes a range of policies and draws attention to the need to expand opportunities for all. In addition to blogging, Anne Marie and her colleagues are present on various social-media platforms such as Facebook, Twitter, and Instagram, where they tirelessly voice their opinions, interact with their followers, and share pictures that speak to their work in the community.

When traveling in the provinces to meet and work with their constituents, women bloggers have taken on the task of training other women in non-urban areas in the use of ICT tools to communicate their needs for development and to counteract cultural and social barriers to their voices. During the course of a recent training for women in the province of Mbuji Mayi, one of the women taking part in the training was nine months pregnant on the afternoon of the training. After completing her first day, she gave birth at 1 a.m. and returned to resume her training at 8 a.m. of the same day. She had just given birth seven hours before, yet she was in the room to complete her ICT training on time.

Connecting Rural and Urban Women

Training rural women on blogging and on ICT tools allows urban women to connect with women living in rural areas via cell phones to identify pertinent issues and share information that is then featured on their blogs. In a country where the Internet penetration rate is just above 3%, the conventional wisdom is to regard Internet access and use as accessible only to the most educated and economically affluent class. While this assumption is not baseless, we need to acknowledge the extent to which access to social media and communication platforms is bridging economic classes in low-income countries where Africans are increasingly using technologies and social media with basic or smart phones across generational, class, and gender lines.

Developing a Crowd-Sourcing Model

After my consulting experience training female political candidates, it became clear that, despite equipping women with the necessary skills, access to financial support was an unmovable obstacle on their paths. Women were over-trained, yet they lacked the basic financial means to support their run for office. It was clear to us that training women was not enough: we needed to address their access to finance.

My co-founder and I decided to launch Fempo, an Internet-based crowd-sourcing and online fundraising space that recruits promising Congolese women within their communities. We raise their visibility by building an online profile for each candidate and directly channel funding raised from individual donors and private-sector partners to support their candidacies. Fempo offers a key solution for the advancement of African women who are remarkably talented and politically capable, yet continue to lag behind their male counterparts in terms of economic and financial power.

Last year, I was able to institute the first Lean In group composed solely of young Congolese women. Following Sheryl Sandberg's Lean In model, we use the Internet to access discussion topics in the free library of online speakers and courses made available to women and men worldwide. The Internet is an important tool that facilitates our learning and growth as we continue to encourage each other and seek leadership advancement in our various fields.

I recently discovered a WhatsApp group called Sukitoko[5] (beautiful hair) composed of young Congolese women. Local standards and thresholds for beauty being heavily Eurocentric in most post-colonial countries, women in the DRC are not foreign to societal pressures around acceptability, colorism and hair. This group started among young women who bonded around their desire to exercise and share tips on natural hair care. The group has grown to become an important platform for information-sharing among young women who build sisterhood and regularly share tips on health, wellness, and job opportunities.

These exciting stories should encourage us to continue to push aggressively for the closing of the digital gender gap around the world, and particularly on the African continent. Across the developing world, nearly 25%[6] fewer women than men have access to the Internet. In Sub-Saharan Africa, the digital gender gap stands at 45% for women (Women and the Web, Dahlberg, 2013).

In the DRC, the under-representation of women and girls in cyber cafes and in the ownership of ICT tools is starkly visible, especially in contrast with the higher levels of economic freedom enjoyed by men. Similarly, boys can simply venture into the neighborhood cyber cafés after school, while girls are burdened with household chores.

As long as women and girls continue to trail in terms of literacy, poverty, and education indicators, the digital gender gap will continue to widen in the developing world. Women's abilities to acquire technological skills and to access the Internet can improve their competitiveness on the job market, so they can start, grow, and promote their businesses, in part by leveraging the power of the Internet, just as their male counterparts do.

As demonstrated in Nigeria, access to technology can enhance women's ability to organize, advocate for public action, and garner regional and international support to their concerns. Amidst the horrific kidnapping of girls in Nigeria, a Nigerian woman, Oby Ezekwelesi launched the #bringbackourgirls slogan, which resulted in women's groups across Africa and worldwide demanding action, forcing governments to pay attention.

Put in the hands of women, the power of the Internet in the DRC can help to determine how women's needs vary across demographics and provinces, and to address the social, political and technology needs of women in a country as large as Western Europe. Closing the digital gender gap will contribute to gender equality, improve the status of women, and continue the work of creating communities in which women's voices are heard and validated in the DRC.

Recommendations

- To build on this progress, women and girls in the DRC need increased financial and technological opportunities to access the Internet and acquire ICT tools.
- Congolese women need educational training and support in order to effectively use Internet tools for development purposes. This can be done by addressing women's needs in technology, starting in schools.

- We need to build technology hubs in the DRC that nurture young people's ingenuity, creativity, and entrepreneurship with a special focus on giving space to women's voices, projects and start-ups.
- Men who unfairly benefit from gender privilege have a critical role to play in challenging discriminatory practices. In the DRC, we need to engage men and boys in gender awareness, starting in families, schools, and the workplace. Beyond awareness, we need to equip men and boys with the necessary cultural, social, and educational tools so they can make the transition from being adversaries to becoming advocates and partners for change alongside women.
- The current and potential contributions of women and men from the DRC will not amount to much until the infrastructure for communication and Internet access improves. In the DRC, as in most countries in the developing world, high cost and low quality remain a challenge for Internet access.

For more information on the women political bloggers in the DRC, please visit the Molongi blog at http://ndimolongi.blogspot.com/

For more information on Anne Marie Makambo, voted best blogger of the year for 2015, please visit: http://kamwanyam.blogspot.com/

For more information on Fempo, please visit www.fempo.net (site under construction. To be launched early June).

7.4 Inviting Multinationals to Support Youth and Technology in Gaza

By Rahilla Zafar

Laila Abudahi received her Fulbright Scholarship in the fall of 2014. She got her master's degree at the University of Washington in Seattle in June 2016. Laila will be joining Palo Alto Networks as a Firmware Engineer. Prior to that, she founded her first start-up in Gaza, where she developed Kinect-based interactive education tools.

How did you decide to study computer science?

I had a very high GPA from high school, and my parents were expecting me to go to medical school. I told them I wanted to study engineering, and they worried I wouldn't be able to find a job. I wanted to prove everyone wrong.

I like building things. I used to work with my father who was always building stuff by hand—tables, shelves, you name it.

What were some of the challenges you encountered in university?

The material provided to students is very outdated. You're completely in your own bubble, not knowing what's happening outside. We think we compete with the university next door, Islamic University or Al-Azhar University, and we don't realize that both of us are just outdated. Also, we don't train students to be curious enough. My own objective while I was an undergraduate was to memorize what the professor was telling me and pass a test.

What was your work experience like in Gaza?

When I wanted to do an internship in Gaza, I worked at a local Internet service provider, where I basically did customer service. I was just answering people's questions, telling them how to reset their router. This is the closest I got to engineering, and it was a bit frustrating.

When I graduated, I got a job at a very big NGO. They were hiring new grads and I was so excited initially, but we spent six months there doing nothing except reading manuals. The organization was paying us, but it didn't trust us enough to do any serious work. It was basically like a charity project that looked so nice on paper because they're creating employment opportunities for new grads, but that's not what it really was.

At that point, I started to take online computer-science courses. I started learning Matlab, a math programming language that is a must for any engineering student.

How did you find out about Gaza Sky Geeks?

I learned about it in 2013. They started when Google first visited Gaza. It was a big deal; it was Gaza's only start-up accelerator. They really sparked a start-up movement and created a culture of entrepreneurship.

The value I see in Gaza Sky Geeks lies in its ability to bring people together. It's a place where people can feel safe and passionate and share ideas. I've met people I would never have met otherwise.

When they started their fundraising campaign last year, I thought, "This needs to work, because this place has to stay open." I don't think we can say,

7.4 Inviting Multinationals to Support Youth and Technology in Gaza

"Gaza Sky Geeks made this a start-up that we can sell now for a billion dollars." Still, I think Gaza Sky Geeks brings tremendous networking opportunities and that it is also having an effect at a cultural level. This is also how I was inspired to bring Microsoft to Gaza.

When did you do that?

I used to be a Microsoft Student Partner when I was an undergraduate. It's like a Microsoft student representative in the universities. You apply, and if you're qualified you're chosen. I was the first one selected as a partner in my university. They were expanding, but they were not in every university in the Middle East and not in Palestine. As part of my responsibility, I was supposed to create a Microsoft Tech Club at my university, and I organized some workshops on coding. It was a small group at first, but then we started to expand, and we created the first Microsoft tech community in Gaza.

I later connected with Naseem Tuffaha, who is the Middle East and North Africa Regional Director at Microsoft. He was interested in coming to see what we were working on, so we organized the first ever Microsoft event in Gaza. This was in 2012, and a team of eight people from Microsoft came to Gaza—these were two of the best days I've ever had. The passion and the hope I saw in the students' eyes were amazing.

It was rewarding to have people from the outside come in and tell you that they see you, they know about you, and that you're not forgotten, despite whatever the political situation that obtains. They say, "We think you're smart, and that you can compete outside." So that was a big boost to us. Students started to develop for the Windows Phone, learning from scratch. I was supposed to be the one who was teaching them, but I became someone who was learning from them. It's about motivation. I don't think we lack intelligence—far from it. We have a lot of smart people, especially hardcore developers, but they need to see a point to what they're doing. I think there are now a lot of freelance developers working on Android and iOS in Gaza, and they're able to make a living through virtual work.

Do you think there's a large opportunity for remote and freelance work?

Mercy Corps has a freelancing academy program. They teach people how to make money online. This freelancing platform can sometimes be very complicated to start with, so the Mercy Corps project is introducing them to how to build trust, how to find clients, how to price their services, and so on. People are actually making a lot of money this way.

Are people becoming developers? Where are they learning how to be good developers?

Everything is online now. The good developers I know, they go online and learn and take courses. And sometimes just stack overflow can make you a good developer.

There is a big power and electricity problem in Gaza, but I think most of the people who freelance get themselves a little power generator just for their laptop and a battery for their router to keep the Internet going. They create a sustainable environment for their work. Of course, they won't be able to control it for the entire household, but they generate a solution that can provide power enough for them to work.

How did you end up in the United States?

I got a Fulbright Scholarship. The selection process takes a year, and they don't tell you that you're a Fulbright Scholar until the moment you pass everything—the interview, the GRE, the TOEFL—there are a lot of steps. You get admissions to multiple universities, and then they tell you whether you're a Fulbright Scholar.

I was supposed to leave in the summer of 2014, and that is when the war started. Being a Fulbright Scholar meant the United States government would help me get out of Gaza. But when the war started, Israel told the Middle East Consulate that they would not allow students from Gaza to leave that summer.

The Consulate called us and said that they didn't think they could do anything. So I had to act on my own. I was like, "I'm not going to let this go." We were supposed to go through Israel and then through Jordan, but that was no longer an option. So I tried to go through Egypt. We applied; there is a quota in the Palestinian embassy in Cairo to allow students from Gaza to leave through Egypt, and we were late, but I applied. And 31 days into the war, I received a call at 6:00 a.m. saying that I got the permit.

The hours before that call were awful. We left our house because our neighbor's house had been threatened with bombing, and we had to evacuate the house at 2:00 a.m. In those moments, my nine-year-old sister saw something that kept her mute, completely silent, for two days.

7.4 Inviting Multinationals to Support Youth and Technology in Gaza

So we evacuated the house. We ran to my cousin's house at the end of the street. I was in my pajamas, and certainly I wasn't planning to travel, I wasn't even thinking about it. I just wanted to survive and I finally went to sleep at 5:00 a.m. after a lot of heavy bombing. At 6:00 a.m., they called my mother and they told her that I got the permit and needed to leave immediately.

My mother woke me up. We left for the border without waking anyone else, because we thought, "What are the chances that I'm going to pass? Am I going to make it?" So I didn't say goodbye to anyone, because they just had just fallen asleep, and why wake them only to come back a couple of hours later?

No one was driving cars in the middle of the war, because they would be a moving target. There was one guy who made an opportunity out of this—he was the only taxi, and people would call him for emergencies and he would charge them double. So we called him and told him we needed to get to the border immediately. There was a cease-fire for two hours—I was just lucky that day.

I didn't think they would actually let me pass, and my mother's phone battery was dying. All I could think of was giving her a charger so that I could have a way of communicating with her if they sent me back. So I didn't even hug her, I just said, "Go to that cafeteria and charge your phone so I can contact you when I'm back." I remember the look on her face. She was asking herself, what if you make it? "What if they let you through?" And they did. And I didn't even hug her goodbye. This is the part that really hurts me every time I think about it.

I had been to the United States once before, in 2013, for a one-month, non-degree program called Economic Empowerment through Entrepreneurship, organized by the University of Michigan. It was three weeks of lectures and studying materials and one week of internship. I had an internship at Dell at that time. That was really amazing. It was my first time in the United States, and I made friends that I still have now. Now I'm at the University of Washington in Seattle, and I'll be graduating with a master's degree this June, after which I can work in the United States for one year, through a post-academic program.

What are your recommendations for creating opportunities in Gaza for youth?

People in Gaza need exposure, motivation, and mentorship, and I think these things are really related. You can't be motivated to learn about something you don't know. In Gaza you're really isolated from everything; that's why I appreciate the work that Gaza Sky Geeks is doing, bringing all of these mentors to Gaza. It is opening doors and letting the outside world know what's going on in Gaza. It's really helping. I think if there were enough people willing to mentor people in Gaza, it would help as there are a lot of people capable of doing a lot of great things.

The English language can sometimes be a barrier for some, but overall people have good language skills. I know that English is the main technical language, especially in computer science and electrical engineering, but if the English material were explained in Arabic, just to get them started, that would be great.

And since we cannot solve the political situation there, freelancing is the real opportunity for people in Gaza. It is not known as a good freelancing destination and that is something we need to work on. If I have an application that I want someone to develop for me, what would make me choose someone from Gaza to build it? I think we need to work on that. If we build our portfolios as a city, as a country, and prove that we can provide high-quality services in the freelancing domain—these services would solve a lot of problems.

I think this is what the Mercy Corps is trying to do, helping people build a trustworthy profile. As a client, just by looking at your profile, I can develop trust in you because I can see what you have built and what projects you have completed. It's doable, but it takes time.

References

[1] http://knowledge.wharton.upenn.edu/article/walk-the-market-tapping-into-africas-900-million-consumers/
[2] http://www.iol.co.za/news/africa-must-produce-technology-1990695
[3] Internet World Stat, http://www.internetworldstats.com/africa.htm#cd
[4] World Bank country profile, http://data.worldbank.org/indicator/IT.NET.USER.P2
[5] Beautiful Hair in the Lingala language
[6] Intel report: Women and the Web, Dalberg 2013

8
Opportunities for Adult Learners

8.1 Queen Rania's Initiative to Provide High-Quality Education through Online Learning

By Rahilla Zafar

Launched by the Queen Rania Foundation (QRF) in May 2014, Edraak is the first non-profit, pan-Arab online education platform leveraging the massive open online courses (MOOC) technology developed by the Harvard-Stanford-MIT consortium "edX."

"The world around us is speeding toward a future where ideas, knowledge, and skills are the founding blocks of prosperity, while we (the Arab world) often drift at the bottom of global rankings in terms of knowledge, and in how and what we produce," said Queen Rania in May 2014, during the official launch of the platform, adding that "transformative shifts usually happen when need and opportunity meet. We desperately need quality education, and online learning is our opportunity."

Her Majesty notes that the Arab world has a chance to acquire capabilities and skills needed to catch up and realize transformative leaps. "We launched Edraak because we realize what we have already missed. It is a way to catch up to a future befitting us, our history, and the message that was sent to us, urging us to read (referencing Islamic history)."

Today, Edraak reaches almost 650,000 learners across the Arab world. It offers free access to open, high-quality learning experiences. Nearly half of its users are women. Its user base includes Egypt, Jordan, Algeria, Morocco, Palestine, Saudi Arabia, Yemen, Tunisia, and Iraq, which are home to some of the world's most global-pressing challenges. The Middle East and North Africa (MENA) region, the most youthful area in the world, has an alarming rate of unemployment among young people in particular, with nearly one out of four Arab youths unemployed, with the average "waiting period" before finding employment at well over three years.

On the other hand, companies across the region have consistently cited a lack of talent and a skills mismatch as two of the main impediments to growth. Employers also face challenges in filtering job applications with current recruitment processes, which can be time consuming and costly.

As a solution, Edraak provides demand-driven courses that are responsive to the region's education and skill gaps, as well as research in areas critical to the region, specifically STEM, entrepreneurship, employability skills, education and teaching, and citizenship education. Most recently, Edraak has introduced an English-language course with the British Council in response to overwhelming demand—the course attracted more than 120,000 learners in its first two weeks.

The Syrian Refugee Support Project

Another new and critical initiative of Edraak is the Syrian Refugee Support Project (SRSP), which leverages Edraak's technology and content to support disadvantaged populations, especially Syrian refugees. SRSP has also prompted the Edraak team to develop technology and content that is most relevant and useful to these populations, thereby mirroring the reality of the Arab world.

The vision of success for the SRSP would be the creation of a cycle that allows refugees to learn discrete skills, use them on the freelance-employment portals, and of course get paid for such work. In the immediate future, Edraak's focus will be on the delivery of access to education.

Adult Learners: Ramia Kiessieh and Naila Mohammad Ali Rajjal

Ramia Kiessieh first learned of Edraak from the news when it launched, and she has been taking courses on the platform ever since. She has a bachelor's degree in banking and finance, and previously worked in a bank before quitting her job to stay at home and raise three children. She feels that the Edraak courses equip her with the skills and knowledge necessary for any potential opportunity that she might encounter should she decide to reenter the workforce, or even launch her own business.

She says courses such as Business Communication have exposed her to skills she had not acquired at university. She also speaks highly of a course she took on child mental health, where she learned how to differentiate between children who are just careless and those with ADHD, for example.

She now feels empowered as the house psychologist, which is significant in a culture where visiting a child psychologist or psychiatrist can be considered taboo.

Ramia is impressed not only with Edraak's capacity to fill skill gaps, but also with the degree to which many experts and so much advice can be found in one place, on a single platform. This includes a course-discussion board, where she gets to meet like-minded people and exchange knowledge and experiences. She says her Edraak education helps her feel less confined by cultural norms or personal circumstances.

Naila Mohammed Ali Rajjal has been a teacher for 15 years, and is impressed with the plethora of learning topics that she can access on Edraak that are relevant to her field. Like Ramia, Naila also recalls that the course on child mental health offered profound insights into how children behave in the classroom, and how to identify and solve problems that teachers face every day. She feels that courses such as these fill a gap for people who can't access education in other ways or who want immediate access to topics that are relevant to them.

Recommendations to Support Scaling Edraak

1. Create public/private partnerships in growth sectors such as healthcare, banking, and ICT to better recruit interns and new employees.
2. Invite the private sector to partner with the SRSP to explore ways to educate and employ refugees.
3. Adapt Edraak so it can meet the needs of parts of the region where the university curriculum is weak or hasn't been updated to offer current technology skills.
4. Adapt blended learning at universities so it can have an impact on professors and learners alike. Professors will be enabled to become more focused on the needs of their students and offer more-personalized learning options.

8.2 Creating a Developer Network for IT and Network Professionals and Software and Hardware Developers

By Rahilla Zafar

Susie Wee is a VP and the CTO of Networked Experiences at Cisco Systems

In 2013, Susie Wee led the launch of Cisco's developer program, DevNet, which provides developers and network and IT professionals with tools, resources, and a learning curriculum necessary to build innovative applications and solutions using Cisco's products. She worked with colleagues Rick Tywoniak, Amanda Whaley, and Parveneh Merat to design and launch the new program. Prior to DevNet, some product teams at Cisco were doing heroics to help developers use their products, but they did not have sufficient resources or support from the company, so Wee began lobbying Cisco to fund a program focused on supporting developers who want to build software using Cisco products, and making sure that these developers are at the center of Cisco's strategy. Wee and her fellow advocates made the argument that, if you want to have a serious software strategy, you need to have a serious developer strategy.

The pitch made its way up the ladder. In 2013, CEO John Chambers (now Cisco's Chairman) made DevNet one of the company's four major initiatives for the year.

"To spearhead the Cisco-wide developer program, the first thing we did was collect all the developer-related information we could find from across the company and put them in one place; we created a web portal that provided one place for developers to go. In doing this, we had to throw out lots of outdated material and create new up-to-date material. This is no small feat in a company as large as Cisco with such a broad product portfolio. We also created tools to help developers code," says Wee. "For example, we created a sandbox, which is a lab that enables developers to get hands-on experience writing apps using Cisco's products, without having to buy expensive equipment themselves. This lets people write apps for enterprise networks, voice and video systems, and contact centers without buying network equipment and unified communication systems. But they do get to code on live networks and collaboration systems that we have running in our sandbox."

Four months later, the team behind DevNet launched Cisco's first developer conference within the company's annual United States conference; they called it the DevNet Zone. They prepared two parallel tracks of talks with coding classes, deep dives into different APIs, panel sessions, and thought leadership sessions with Cisco executives and guest speakers. They hosted a 24-hour hackathon with live kit so that people could hack on Cisco software and hardware. They also created "learning labs"—self-paced learning labs where people could get hands-on while they learn; they could have access to a running version of a live network with Cisco's software defined networking solution, then call an API to it and get information back. They can make API

8.2 Creating a Developer Network for IT and Network Professionals

calls like "get network devices" or "set QoS policy." This allows network and IT professionals to learn to code in the context of the real work that they do.

"When we built this, we didn't know how many people would really be interested. We didn't know if network operators who were getting their certifications were ready to learn to code. So we just got this whole party ready, in the same space where they hold the Google I/O and Dreamforce developer conferences, and wondered what would happen. We made the DevNet Zone as accessible as possible—practically free. The first day we opened people came and sat down at the learning labs and just stayed there. Then I heard some of them call their work colleagues back in the office and say, 'Hey, you guys have to come up here to San Francisco. Cisco is doing this really cool thing.'"

Almost immediately, the learning labs filled up. By the second day, the conference was packed from wall to wall; the interest exceeding expectations. John Chambers came by and saw the engagement. He called people on his staff and told them to come over to see the DevNet Zone. After that, DevNet became established in Cisco not as a one year strategic initiative, but as Cisco's developer program.

"When we started DevNet, we thought it was going to be a standard developer program to help people code," says Wee. "We offered a Coding 101 class to help network operators—super-smart people who run the world's mission-critical networks but for whom coding hasn't been part of their day jobs—to understand the shift to software-defined networking. It turns out that there was tremendous thirst for the value we were providing in learning. So then our mission evolved from providing developer resources to grow a developer ecosystem of open solutions to also helping developers learn and grow their careers."

David Ward, Cisco's Chief Architect, supported Wee in incubating the program. "The point of having a developer community is not only to get more people to understand the different layers of the platform, it's to enable them to educate themselves so they can grasp new opportunities. It's also for developers of another kind, who want to express themselves as engineers and accomplish something without having to learn everything there is to know. We are currently asking people to understand the complete history of the Internet, the construction of the Internet, the protocols of the Internet, the layers of everything from top to bottom to be able to express themselves. DevNet has the opportunity to short-circuit that and say, 'Here's what you need to know to work in a cloud or deliver this service or publish this content.'"

Ward sees this as an exciting opportunity, as developer communities that are focused on networking are in nascent stages, simply because it's one of the last high-tech industries to adopt API, platform, orchestration and programmatic capabilities. "There are many services being delivered in an uncoordinated way as different platforms and different sets of APIs. We're at the very beginning of building out that developer community and the community of people who want to build their businesses on top of our platforms," he says.

DevNet's Next Frontier

The DevNet community has quickly grown from 50,000 to 370,000 developers. This year, the DevNet Women in Tech Community was launched to support women as they continue to develop and build on their technical skills industry-wide.

"We still have a lot more work to do," says Wee, "and we're constantly improving and adapting. We're like a start-up within Cisco, but now we're in a good place and ready to help more people. One of the areas that we really want to grow in is what I'm calling Communities of Interest—providing technical training, events, and a supportive environment for different communities, such as women-in-tech."

The DevNet team is excited to leverage the learning labs and materials for the benefit of more people at various levels of skill and experience who really want to learn about and contribute to these areas.

"It's scary to participate in a hackathon if you're new at it," says Wee. "We're hoping that we can help people learn some of the needed skills and gain the confidence to participate in a real hackathon or coding event. There are so many people who are smart and who want to contribute something, but the technology has changed so quickly: how can you get your foot in the door? You don't have to do it alone—you can do it within a community. That's where you get the tools and the support that you need."

As a 45-year-old woman who has just had her first child, Wee understands how hard it is to keep up and stay relevant when life events require time away from the job. She now has a special appreciation for the value of online learning. DevNet provides abundant online resources; the learning labs, Coding 101 classes, and sandbox can all be accessed online from anywhere.

"What's nice about this new world of software is that you can actually do it in a less-structured way that works with your lifestyle. There is a world of hackathons, a world of software development to which you can contribute. Everyone needs a coder, so you can have a career that doesn't require a typical, structured job. Some of my college classmates who are MIT graduates, are super-smart women who decided to take time off to be with their kids and not work professionally. There's an amazing potential workforce there, because they're some of the smartest people that I know. If given a way to participate, they can definitely make the world a better place. I want to help make this happen."

8.3 Are We Too Old to Be of Value?

By Laurie Cantileno

For women over 50, ageism is yet another type of discrimination, whether it's in career advancement, or for those women coming back to work, or looking for another job or even starting a business of their own. Elizabeth Isele, Senior Fellow in Social Innovation at Babson College, and serial senior entrepreneur at the age of 70, launched The Global Institute for Experienced Entrepreneurship (GIEE). In addition to being its CEO, she is also co-founder of SeniorEntrepreneurshipWorks.org, co-founder of eProvStudio .com, founder of SavvySeniorsWork.com, and co-creator with Participant Media of the 2015 global summit series on Senior and Multigenerational Entrepreneurship. After a long publishing career as an award-winning editor, Elizabeth now brings generations together to create businesses and aggregates research and public policy to support people over 50 around the world who are starting businesses.

The research Elizabeth is developing with universities has separate components, but all are focused on documenting the impact of senior and multi-generational entrepreneurship. Some of the research is gender-specific, such as the work she does with the EMEIA (Europe, Middle East, India and Asia) initiative at Ernst and Young (EY). EY is committed to bridging the skills gaps through a program called WFF (Women Fast Forward) to accelerate the 117 years the World Economic Forum predicts it will take to close the global gender gap. That's something Elizabeth is gathering data on for EY and Chatham House for the upcoming G20 Summit in China in September 2016. She is also helping EY study new ways to better position women to take advantage of opportunities throughout their entire EY career arc, as well as providing valuable data about women who start businesses of their own, and implementing her Experience Incubator® as one of EY's top five global-growth strategies for 2016.

Elizabeth explains: "It's very exciting, but it can also be a bit daunting because the impact of the 50+ entrepreneur has the potential to be socially and economically transformative in ways that directly affect prosperity for people of all ages around the world. For too long, governments and corporations have seen the super growth of this demographic as a pending disaster instead of the opportunity it represents. The fact is, we have the largest, most experienced talent pool in the history of humankind, and as I travel the world raising awareness of this opportunity, people keep introducing me as the expert in this arena. I look out at the audience and say, the only reason I'm called the expert is because for years, I was the only one talking about this subject."

There were two catalytic moments for Elizabeth. The first was way back in 2009, when she felt intuitively that older women wanted to continue working, wanted to remain in the workforce, and she thought, "I need to test that idea." So she asked her webmaster how to create a blog, and her webmaster said, "Well, you could write whatever you want because nobody's going to read it."

Elizabeth, of course, took that as the ultimate challenge and published the blog. She said, "When I realized that her comment actually gave me a lot of latitude to write whatever I wanted, I set about writing a blog just as I would talk with individuals. It was fun rather than a dry bit of focus-group research, and contrary to my webmaster's dire predictions, I had responses from every continent in the world, except for Antarctica, within six weeks. I knew then that this was not some intuitive fluke. This is real." The blog is SavvySeniorsWork.com, and it has all different kinds of tips, information and insights for both men and women who wish to continue being employed after the age of 50.

She asked the president of the Philanthropy Roundtable, where she was Director of Economic Opportunity Programs, "Why are we doing nothing for vulnerable people over 50?" He said, "I really don't think our donors would care about that."

She responded, "But that strikes me as strange because most of our donors are over 50 and most of them were also entrepreneurs." He said, "Well, yes, but ......" so, again undaunted by a challenge, she asked him, "Well, do you mind if I research that while I'm looking for all these other programs for our donors to support?" He said, "No, no, no." He just assumed it would be nothing. And sure enough, when she started researching it, it was nothing and there were no programs out there.

So she thought, "Well, I've got to do something about this." That's when she resigned from that organization and started Senior Entrepreneurship Works

8.3 Are We Too Old to Be of Value? 225

in 2012. That was the genesis of it, and Senior Entrepreneurship Works has evolved into the Global Institute for Experienced Entrepreneurship, because of the interest in it from around the world. Elizabeth changed the title to Experienced because she really wanted to draw on the intergenerational component to optimize the technology and social-media experience of younger people and the life-work experience of older individuals, because she thought, "That's where it's really going to be catalytic and that's where it's truly going to boost prosperity for people of all ages."

Ageism and Elizabeth Hit the Big Screen

So, how did this then evolve into an advisory role for Hollywood? Elizabeth explains, "In my very early workshops I was using *The Best Exotic Marigold Hotel* as an incredible example of intergenerational entrepreneurship and how that works. People knew and loved the movie and got the point. Then, one day she had a call from the office of Jeff Skoll, the movie producer. His company, Participant Media, produces movies with a social purpose—the latest of which is the Academy Award winning Best Picture, *Spotlight*. They said, "We understand you've been using our movie in your workshops." I, of course, with my publishing background, almost had a heart attack and said, "We were but really ... I did not mean to create any infringement of copyright. It's just an extraordinary example of how this works." They said, "No, we're not here to criticize that. In fact, we would like to ask you if you'd be interested in creating the social action campaign for the sequel to *The Best Exotic Marigold Hotel*."

Elizabeth said, "As people have come to know, the older I get, the bolder I get, so I said well, I'd really be interested if we could do something more proactive than just giving people a prize for starting a business after 50, which they had done in the past with the first movie."

They asked, "What do you have in mind?" and Elizabeth said, "I would like to host a series of summits to bring together cross-sector leaders in different parts of the world to amplify the voice of senior entrepreneurs and build a cross-sectoral ecosystem that understands the economic impact and support needed from financial institutions and governments to drive this new economic engine." Her idea was based on a similar, enormously successful summit she had hosted in Washington DC with the Gerontological Society of America in 2012.

In the summits, a group of 20 to 25 pro-active, powerful thought leaders come together from different sectors, including finance, education, research and public policy to understand the opportunity that senior entrepreneurship provides and identify what they can do to support it. Elizabeth told Participant she'd like to replicate the summits around the world as their social-action campaign. She suggested they invite people to a special pre-release showing of the movie, and then host a mini forum afterwards. They loved the idea. Top-level people came whenever they were invited, and that's how Elizabeth's organization extended its global outreach so quickly.

She said, "Governments and corporations around the world are trying to figure out what to do with the bulging demographic created by our newly extended longevity." Elizabeth is changing the perception that this is an impending tsunami, saying, "This is far from doom. This is a huge social and economic opportunity for people locally and globally."

Company Loyalty a Thing of the Past

It's a new working world. Young people entering the workforce today eschew their parents' and grandparents' ideas of having one career with one company until they're ready to retire into a rocking chair. They are thinking in terms of a three-year stint with one firm at best. Corporations that are spending huge sums recruiting and training top talent ignore this at their peril.

This is where Elizabeth's Experience Incubator® comes in. The incubators are designed to catalyze multi-generational experience to boost employee engagement and knowledge transfer through entrepreneurial thinking and acting. In the workshops, young people share their social-media and technology experience, and older employees share their work and life experience. "Most corporations," she says, "have two problems: one is that younger people coming in have no intention of staying after you've invested in them, and a second is that you're pushing the older employee who's got years of knowledge and experience into retirement without any knowledge transfer process. This is especially impractical when those older, experienced employees have no wish to retire."

She continues, "This is compounded by the fact that the number of people coming into the organization is shrinking." This is not something that you should think about in a leisurely fashion. This is coming to the crisis stage. Corporations should be pro-actively rebooting people across their entire career arc. The statistics on employee engagement are absolutely horrific at any age. When we're talking about just 10% of employees worldwide completely engaged in their work, it has a heavy impact on the employees' psyches and takes a heavy toll on the firm's productivity. To make matters worse, to that

10% of staff who are disengaged, you can add those employees who typically drop out five years prior to retirement. The good news: research demonstrates that those companies that nurture engagement in a multi-generational workforce increase their productivity by at least 20%.

You really get the generative action of everybody's experience when you show young people coming in, mid-level people, and even pre-retirees, what other opportunities exist that you as a corporation are going to support. These may not represent a traditional linear career path. Exciting new paths can be created tangentially to reboot or redirect engagement—and that will, in turn, enhance retention. Elizabeth says, "Every job requires an entrepreneur. You have to think entrepreneurially about your work no matter what level you're at." Elizabeth calls it an ROC (return on career). It's a different kind of return on investment for both employees and corporations.

I Want to Be My Own Boss Now

The data from Kaufman shows that over 80% of people over 50 who want to start their own business do it because they want to do it. It's not because they have to do it. It's only 7% of people who say financial reasons are the only reason they're starting a business after 50.

Elizabeth says, "People over 50 are fiercely driven to remain relevant, and they want purpose in their lives. There's no blueprint for these additional 20 to 30 years that we have been given. We're all entrepreneurs in that sense, trying to figure out what to do with our extended longevity. People approach building a business of their own for different reasons. Some are acting on a life-long dream they've had to repress to meet financial responsibilities. Many are more than ready to find a way to be their own boss after working for others for years. Others are keen to apply their talents and experience to give back by solving a problem of value to society and their community.

"The biggest challenge, and especially for women in starting a business of their own," Elizabeth says, "is a lack of confidence. That's the number one challenge for women. Women do not see their life experience as entrepreneurial."

Elizabeth tells people, "You most probably have been operating as an entrepreneur for your entire life, but you've just never identified as one." She says, "The word 'entrepreneur' can be very intimidating. So we try to get people to think of entrepreneurship as how you solve daily problems in your life instead of equating it only with what the likes of Steve Jobs or Bill Gates have done." She said she often uses her own experiences as an example: "I was a single mother of four, each of whom had some kind of sporting event starting approximately at the same time on Saturday mornings. I was one driver with one car, and I had to apply every ounce of entrepreneurial thinking to get them to their events on time, allow a few minutes to cheer them on, and remember to pick them up on time."

Elizabeth noted the most popular exercise in her workshops is called "Decoding Your Entrepreneurial History." "We go back decade by decade to help people see how they have been acting entrepreneurially all their lives and identify the patterns and strategies they applied to make each of these endeavors a success. It's empowering because it takes away the intimidation of entrepreneurship."

Judi Hendersen had been laid off from her technology job and couldn't find a new one during the recession. She took a break from reviewing job postings on the Internet to scout out some Tina Turner concert tickets. She unexpectedly found someone online selling mannequins, and said, "Oh, I've always wanted one of those for my garden." She called the seller who said, "Well, actually I'm going out of business, so you'll have to take all 50 if you really want it." Thrown for a moment, she thought, "What am I going to do with 50 mannequins?" Then he said, "That's it. Take it or leave it." She said, "Okay, I'll take them." She picked up the mannequins and brought them home to her garage in Oakland, California. Then she said, "What am I going to do with all of this?"

To cut a long story short, Judi went to a younger colleague who was very adept at social media, especially social-media marketing. Together, the two of them set up "Mannequin Madness," an online business where she started selling her 49 revitalized mannequins (she saved one for her garden). Today, she's grown that business from 50 mannequins into thousands, and she sells them around the world.

This is also a very green business because she is saving tons of material from landfills. Judi even won an award from the EPA (Environmental Protection Agency) for her efforts.

The United Kingdom, Ireland, France, Israel, Spain, and other parts of the EU are far ahead of the United States in the senior-entrepreneurship arena. The EU has cited Senior Enterprise as a key strategy for achieving its 2020 economic-prosperity goals. And research in Spain has documented that, for every euro the government spends to mitigate what they call the "retirement syndrome," they will receive 129 euros in return. Elizabeth notes that, "No matter what, you have to understand the importance of technology in your entrepreneurship. But that being said, if it is something you understand and you understand the need for it, you also do not have to do it alone. You can bring in somebody who just absolutely lives on technology and has that incredible expertise as a partner, as a colleague to help you build your business."

8.4 Open Source: Shifting the Corporate Innovation Model (and Leveling the Playing Field)

By Lauren Cooney, Senior Director of open source strategy at Cisco Systems
Twitter: @lcooney

Lauren Cooney is the executive lead for open source at Cisco Systems, where she drives strategy, community, product and technologies around open source efforts. Cooney has over 15 years of open source experience across BEA, IBM, Microsoft and Cisco. She is passionate about open technology and women in tech efforts/STEAM (Science Technology, Engineering Art and Math) and resides in the San Francisco Bay Area.

"The more we create our products as open platforms to which people from the outside can contribute, or build on, or roll into their own products, the more we will open doors for women or anyone looking to transition. It helps create lower barriers to entry. For example, one doesn't have to learn IOS in order to program a router. One can use easily accessible APIs just as one may have done in other jobs. This approach also provides a community of people who are developing on the same platform, and that is a potential avenue that creates an opening to encourage more women to participate in the development of new products." —Alissa Cooper Cisco Systems

A Bit of Background

My first exposure to an open source project was fifteen years ago, when I was just 23. I was working for BEA Systems, and we decided to open source some of our code into the Apache Software Foundation. (For the grammatically aware—yes, "open source" can be used as a noun, adjective, or verb.) That project meant a lot to me, because it was the one that drove me to code for the first time. I was leading a marketing team, but my engineering counterparts insisted that, in order to be successful at *marketing* an open-source framework, I'd have to learn how to *code* in one. Keep in mind, this was long before code camps, and we didn't yet have the bevy of reference sites and code sample libraries that we have today. They just threw me into the deep end: "Here's a Java book. There's a CD-ROM in the back (if the last guy didn't swipe it) and give me a Hello World app by tomorrow."

So off I went and low and behold, I learned how to write a few apps. Granted I was using an IDE (Integrated Development Environment; read: drag and drop, compile and run), but it set me on my way to learning how to build applications in a number of different languages. It also exposed me to the open source "movement," which BEA was looking to build and grow.

After BEA, I continued to build upon my successes at IBM, Juniper, and Microsoft, where I worked on a number of open-source projects, including the establishment of Microsoft's first Open Source Foundation. I've learned so much from the men and women at the forefront of the open-source community over the years, and the momentum I have built in those circles has most recently brought me to Cisco, where I'm now working to create an even stronger drive for our customers, developers, and partners interested in open source.

Open Source Is a Great Buzzword, but What Does It Mean?

Wikipedia defines open source as "computer software with source code made available with a license in which the copyright holder provides the rights to study, change, and distribute the software to anyone and for any purpose. Open-source software may be developed in a collaborative public manner."

More simply put, open source is a license and model of development that is typically decentralized and executed via a community of peers rather than in a traditional, closed corporate-development team. The code is open for contributions, and is distributed to the general public for use and modification from its original design. That means developers can download, modify, and/or re-publish their code back into the community, or (if the license allows) keep it in-house for use in proprietary products.

Shifting the Corporate-Innovation Model

Historically, software companies have thrived on proprietary products: products built for profit that cannot be modified or openly distributed to the developer community. That is changing. Corporations are now, more than ever, contributing to and promoting the usage of open source for rapid innovation. More importantly, though, they are beginning to see open source as a competitive advantage, or even a competitive necessity in some industries. Microsoft, for example, is now running Linux on the Azure Cloud—a stark contrast from their past competition with the popular open-source operating system. This changing dynamic has given incredible momentum to the open-source movement. It has shifted the power dynamic in favor of developers, and the innovation model toward a more open, collaborative one. It has also helped to build bridges across domains, geographies, and even intra-company silos—perhaps the most difficult barriers to break.

There is another interesting shift happening though. We are seeing more companies than ever before open-sourcing entire libraries of their most important software assets. Cisco, for example, recently worked with the Linux Foundation to establish and seed the FD.io project (aka Fast Data Project, focused on data IO speed and efficiency for more flexible and scalable networks and storage), and they have more open-source projects on the way.

Why open-source something that could be a proprietary crown jewel? Simply put, the open source user and developer ecosystem is too big to ignore. Users and developers alike now expect products to be free and open to use, modify, and build upon. Releasing a project to open source is perhaps one of the fastest ways to innovate, gain developer traction, and move the industry forward at a pace that large corporations have been struggling to match.

Leveling the Playing Field

The Open Source Initiative, or OSI, states that open-source software doesn't just mean access to source code, but encompasses a broader philosophy, including tenets such as free redistribution, derived works, author source code integrity, technology neutrality, and *no discrimination against persons or groups or fields of endeavor.*[1]

Corporate innovation has historically happened in small groups or closed environments (labs or R&D centers), and developers have been at the mercy of their corporate cultures. In the open-source community, things happen publicly with larger groups of developers who do not see themselves as tied to a particular corporation, country, language, or other stratum. Developers have flocked to this model because they can innovate faster than they could alone or in a more isolated environment. They can grow their careers more independently and their work is judged by its quality and adoption—not by corporate politics, gender, race, or other subjective means. Some projects that exemplify this that we use every day are solutions such as WordPress, the website and blogging technology which is built in open source; Mozilla Firefox, the second most popular web browser in the world; and the Apache Web Server, which runs the majority of websites on the Internet today.

In short, the open-source community allows for a wider net to be cast—a merit-based net that invites new ideas, disruptive innovations, and where code is king. What better environment is there for women than one in which ideas and contributions are valued based on their own attributes—not on those of their creators?

As the open-source momentum continues to grow, development and career opportunities for women are expanding, as is the recognition that women are achieving in this space. Red Hat routinely recognizes Women in Open Source, both in industry and academia.[2] CIO Magazine recently put out an article titled "Ones to Watch: Influential Women in Open Source," which highlighted some amazing achievements by women in open source.[3] Open-source conferences, such as those held by the Linux Foundation, are actively recruiting women to be on open-source panels and speak at their events. The Grace Hopper Celebration of Women in Computing (GHC) conference now hosts an annual open-source hackathon. The OpenStack Summit also hosts a "Women of OpenStack" event each year at their annual conference to encourage more women to collaborate on the project. O'Reilly's OSCON (Open Source Conference) 2015 included more women speakers and hackers than ever before. I find this an amazing shift from the days in which I was one of the five to ten women routinely attending these open source shows.

There are also more women now engaged in and leading key open-source initiatives, including Mitchell Baker, Chairwoman of the Mozilla Foundation and Corporation; Danese Cooper, leader of PayPal's open source initiatives and a board member to several open source projects; Sarah Novotny, Kubernetes leader for Google (and former open-source leader at Nginx); Stormy Peters, who leads several efforts for CloudFoundry.org; and many others who are breaking ground by leading projects, evangelizing, and—most importantly—contributing code. The community is just that—a community—and each and every person, male or female, has an equal opportunity to lead and contribute.

Is that enough? No, but it's a start. I have made it a personal mission to evangelize, inspire, and encourage more women—and more importantly, *young* women—to learn and love code. I speak to these women at every opportunity, telling my story, and trying to give them a baseline understanding of how the open-source community is changing the world, and how they too can be a part of this movement. I hope that you'll join us. Pick up a book, attend a conference or a hackfest. It's never too early, and it's never too late.

References

[1] https://opensource.org/osd
[2] https://www.redhat.com/en/about/women-in-open-source
[3] http://www.cio.com/article/3066864/open-source-tools/ones-to-watch-influentialwomen-in-open-source.html

PART V

Breaking the Glass Ceiling: A Generation of Women Forging ahead into Technology Leadership

9

Stories of Resilience, Perseverance, and Staying on Top of Technology Trends

"In all the spheres of society where there is tremendous gender inequality, the one area where change can happen overnight is on-screen. It's going to take forever to get Congress to be half women, but there could be a half-women Congress tomorrow in the next movie somebody makes." —Geena Davis

9.1 Introduction

Something else the women highlighted in this chapter have in common is that they paved the way for many others. The women we feature stayed resilient and built trust among team members who initially questioned their abilities. They also paid attention to technology trends and continued to build new skills—and that helped them to continue to advance in their careers.

9.2 Breaking the Glass Ceiling in Saudi Arabia

By Rahilla Zafar

"As the first woman in IT at Samba and the Saudi Stock Exchange, and from my experience in changing a culture within those environments, I was inspired to do more to support more women in IT."

Deema AlYahya is based in Riyadh, where she is Director of Developer Experience & Evangelism for Microsoft.

Growing up, Deema AlYahya was inspired by watching her father work on gadgets and computers. From the age of 12, she knew she wanted to work in technology. She dreamed of studying at MIT and returning to Saudi Arabia with a unique degree—such as one in artificial intelligence—that was not offered in the Kingdom's universities at the time. AlYahya got engaged shortly after graduating high school, however, and that meant that she could study only within Saudi Arabia, where degree programs in information technology for women were restricted to software engineering.

AlYahya never allowed herself to feel limited by her circumstances, however, and her father's advice that one can find knowledge anywhere was central to how she approached her education and later her work. While at King Saud University, AlYahya began taking online courses in artificial intelligence, which eventually became the focus of her graduate studies. Upon graduating, the options were once again limited for women in technology; most women opted to work in female-friendly environments such as universities, schools, or hospitals. AlYahya joined the financial sector, working for Samba Financial Group, and insisted on applying to their IT department. She ended up being the first female software engineer working with a group of expat programmers. She admits to feeling awkward at first, with no one there to help or mentor her. But she noticed that many of her colleagues were constantly studying and earning certifications, and AlYahya followed suit, earning five technical certifications.

Saudi Stock Exchange: Building Trust

At Samba, one of the areas that fascinated AlYahya was a project for the website of Tadawul, the newly launched Saudi Stock Exchange. It was 2004, and AlYahya set her sights on being the first woman to work for Tadawul; Saudi leader King Abdullah had just given permission for women to work in all professional sectors. Despite this new law, when AlYahya applied she was told they were not ready for female employees. AlYahya refused to give up and did research on other stock-exchange websites around the world. Based on this research, she submitted a proposal to Tadawul on how to develop a local website that would be on a par with international stock exchanges. Shortly after submitting her proposal she received a phone call for an interview, but later learned it was only because the Capital Market Authority in Saudi Arabia had forced them to comply with the new law of the King and recruit women.

Once hired, AlYahya was given a $4'' \times 3''$ room with a high-end PC and an Internet connection. She wasn't allowed to leave her office or go to the male section. If she wanted to meet with a colleague, her manager had to serve as her male guardian. She began to think, "Okay, now what should I do? I can't stay like *this*." She wanted to contribute to the change needed in Saudi Arabia so there would be a better tomorrow for her daughter, sisters, and nieces. But she wanted to be careful. There were too many stories of women pushing aggressively for cultural change that had had counterproductive results. AlYahya decided the pathway to a more ideal work situation was to gain trust among her colleagues.

How could she gain the trust of her team so they would see her as integral to their growth and success? She thought of her manager. Is there a way she could work with him without him feeling threatened that she might take his place one day? She revisited the same research that she had sent as part of her application process, which had never been read, and sent it to her manager, telling him that if he liked it and wanted to present it to the leadership, he could put his own name on it.

Her manager did just that, and received a lot of questions he couldn't answer from the leadership. He called for her help. AlYahya answered these questions and continued to insist on not being named. The CTO was impressed and decided to develop the ideas in the proposal. The development team enlisted her help, leveraging her experience at Samba. She told them they didn't need to give her credit. The site became a huge success, and her team was praised for their efforts. Most importantly, her plan had worked, and she had won the trust of her team.

During her six years at Tadawul, AlYahya helped the HR team launch a diversity program that recruited 25 women, including in management roles. She was also tasked with learning a new platform—Microsoft SharePoint—that no one at the company yet knew how to use. Given no materials or instructions, AlYahya used her personal vacation time and funds to travel to London and receive training at the Microsoft Institute, earning her certification in the platform. Subsequently, when the Tadawul Board of Directors requested a collaboration platform, she was given the project. She knew she could handle this on her own, so she got the requirements and built out the collaboration platform using SharePoint. She was asked to present the tool to the board members, most of whom are now government ministers. Because of this presentation and her knowledge of SharePoint, AlYahya was offered a job in the Ministry of Foreign Affairs.

The Foreign Ministry's IT Department

AlYahya proudly left the Saudi Stock Exchange. "As the first female in IT at Samba and the Saudi Stock Exchange, and from my experience in changing a culture within those environments, I was inspired to do more to support women in IT," says AlYahya. She became the first female manager in the

Ministry's IT department, managing a team that oversaw all of their online platforms, Internet, and social media. She again found herself in a unique work environment, where male and female departments were separated. She was not allowed to go to the male section despite being their manager and instead relied on a virtual conference room.

Colleagues warned that she wouldn't be accepted as a manager, so she decided to deploy the same strategy she used at Tadawul and worked to build their trust. She asked her team to give her three months, and if they weren't happy she'd personally support their move. After three months as their manager, none of them wanted to leave.

During her time at the Ministry, AlYahya felt not only that she had created a more accepting work environment for women, but also that she had created a culture of productivity during downtime. She encouraged innovative thinking and challenged her team to do things differently. Others within the Ministry began to take notice that her employees always actively sought more work, which was an impressive cultural shift. Young women would seek out her mentorship and share their ideas with her. She invited other women to work in her otherwise private office to support more collaboration. By the time she left the Ministry they restructured the entire female section to remove private offices altogether and thus continue the collaborative culture she had instilled. AlYahya recalls her time in the Ministry fondly and felt she really made a difference there.

Microsoft: Continued Trust Building

In 2012, when AlYahya was 31, Microsoft offered her a position that she was delighted to accept. While Microsoft in Saudi Arabia had women on staff, none were in senior roles. AlYahya became their first female Saudi manager, overseeing a team of 20 senior architects, consultants, and project managers. For most of her team, it was the first time they had both a woman and a younger person managing them. She felt the team she managed initially underestimated her and assumed she was hired because of the King's call for diversity in the workplace. As in her previous roles, AlYahya embraced this challenge. Under her management, the productivity of the team increased. More recognition, rewards and bonuses were provided for their hard work than under previous management. She also took a serious approach to work-life balance after seeing how many team members were burned out after months spent on offsite projects, and she adjusted their schedules accordingly. Under her leadership, her department's health-index score, which measures how happy people are in their work environment, increased by 30%. AlYahya was soon promoted, and became the first Saudi female executive at Microsoft Saudi Arabia, overseeing the developer experience.

Supporting Women in IT

AlYahya also noticed that while women in IT in Saudi Arabia are getting their MBAs and growing into COOs and CEOs, few become CTOs or CIOs. Even Princess Nourah Bint Abdul Rahman University, which offers one of the country's largest IT programs for women, has men serving in CIO and CTO roles. In response to this trend, AlYahya created an initiative to empower females in IT leadership. Taking advantage of such programs as Microsoft Spark, which gives employees the freedom to give back to their communities by investing in proposals they submit, she supported women in the IT sector to become leaders through training sessions in soft skills across the country. These "mini MBAs" seek to increase confidence and the level of professionalism among women with IT aspirations, and provide communication, negotiation and financial and technical training. As a next step, she's now looking to bring this initiative to school-age girls and expose them to the IT industry at a young age.

9.3 Building the Foundation for Innovation

By Rahilla Zafar

"If I had waited for mentors who looked like me, I'd still be waiting."

Michelle Lee is Undersecretary of Commerce at the United States Patent and Trademark Office.

The first woman to head the United States Patent and Trademark Office (USPTO) in its 226-year history, Michelle Lee had a childhood that sounds similar to that of Apple founder Steve Jobs. She grew up in Silicon Valley. Her first mentor was her father, who kept their garage stocked with transistors, resistors, circuit diagrams, wire strippers, and a soldering iron. Her father built the family's TV set, and her own curiosity was inspired by watching him take things apart, tinker, and create things. Together they built a hand-held radio, and growing up on a street where most of her friends had a parent who was also an engineer, she thought such activities were what every girl did. However, as she grew older, she realized this was not at all the case.

In high school, as Lee moved into calculus and advanced calculus, she found that she was one of a small number of female students. Later, at MIT, where she earned bachelor's and master's degrees in electrical engineering and computer science, it was the same story. She was one of even fewer women who, as a graduate student, worked as a researcher in the MIT Artificial Intelligence Lab. Lee learned early on not to limit herself by seeking out only those mentors who looked like her. "If I had waited for mentors who looked like me, I'd still be waiting," she says.

Recognizing an Increasingly Competitive Global Environment

Now in her current role, among Lee's many priorities is an emphasis on advocating for women and girls so they can have the exposure and opportunities available in STEM fields. She's quick to add that it's not because of her gender that she's advocating for women and girls, but because she's a person who understands that in order for the United States to stay competitive within the global economy, it needs to empower all of its workforce. She notes that, at the broadest level, the mission of the USPTO is to promote American innovation, and to her that means across all geographic regions of the country and demographics.

According to a recent Department of Commerce report, STEM jobs are being created at three times the rate of non-STEM positions. And according to the American Association of University Women, the United States will need almost two million new engineering and computing professionals in the next seven years. While STEM jobs pay higher than average, two of the professional areas in which women are the least well represented are computer science and electrical engineering, which are fields in which Lee studied. As the first woman to lead the USPTO, Lee inquired how many women were listed as inventors on a United States patent. One study that looked over a 30-year time period determined that less than 15% of the United States-based inventors listed on a patent were women.

"The low number clearly leaves room for improvement. It is an economic imperative, as we cannot afford to leave behind any inventor and entrepreneur. We never know who will invent a technology that will revolutionize the world or the way in which we live. And in an increasingly competitive global economy, we need to take advantage of *all* of our talent," she says.

Lee reflected further: "As I sit across the table from my counterparts at meetings around the world, it's like the United Nations, where you're at this huge U-shaped table and people sit by country with their national flags. It's very interesting to see who sits in the first row, who in the second, and so on. Are there any women in their delegations, and where are they sitting? I will say there are certainly differences. The United States is probably not the worst but it's not the best either."

Addressing the Gaps in STEM Education and Opportunities in the United States

Lee says there is still room for improvement in terms of both recruiting women to enter the STEM and computing fields, retaining them, and having them rise through the ranks and enter leadership roles. As the head of America's innovation agency, she sees tremendous levels of inventiveness, innovation, and entrepreneurship in the United States. Though technologies have evolved over time, from inventions focused on agricultural developments, to mechanical engineering, to the personal computer that propelled software development, to the Internet and mobile, not to mention incredible innovations in the biotech and pharmaceutical industries, she sees the fruits of an incredibly innovative society passing through her office. But she would like all segments of society to participate directly and equally in what, to her, appears to be a very bright future based on innovation.

To this end, she has launched and supported numerous programs at the USPTO, such as Camp Invention, a partnership with the non-profit Invent Now. Through Camp Invention, each year more than 100,000 elementary-school-aged kids in all 50 states participate in a week-long summer enrichment program. The kids get hands-on experience in how to design, prototype, build, test, and refine a specific device. The USPTO has focused in recent years on expanding its reach into underprivileged communities, with the National Inventors Hall of Fame offering scholarships for students. Another partnership is with the Girl Scouts of America, which led to the creation of a patch on intellectual property (IP) and innovation. To earn an IP patch, the girls learn about the fundamentals of patents, trademarks, and copyrights, then put their innovative spirits to work on creating something. Lee recalls that, when she was a Girl Scout, the patches were for first aid and sewing.

"I want more girls—and boys—to grow up wanting to be inventors and entrepreneurs. That's across demographics, all geographic regions of the United States, not just the East Coast or West Coast but in the middle of our country as well," she says, recalling that not all families had scientist role models like hers. According to Lee, the United States does a decent job in providing textbook training on math and science, but for her it's important that youngsters are also inspired and incentivized to apply that knowledge to create new designs, innovations, products and services. Lee traveled to Iowa recently and worked with officials there on how the USPTO could bring its resources to help support STEM education. Under the Department of Commerce, there's a priority to support more hubs of innovation. A number of factors must come together to incentivize these hubs, including educational training, financial capital, and great universities where research and ideas can be developed. Also, a hub must be in a part of the country where faculty and students want to settle. A talented labor force and affordable housing are also critical, and creating these types of innovation hubs has been a priority of the Obama Administration. Lee's office has provided support for these efforts across the country, including through education and outreach to train teachers on intellectual property, invention and entrepreneurship, as well as programs such as Camp Invention and the Girl Scout IP patch that spark youngsters' interest in invention, making and entrepreneurship.

The National Inventors Hall of Fame, a USPTO education partner, is best known for recognizing luminaries in the field who came up with such inventions as the Post It^{TM}, digital-camera technology, and 3D printing, as well as many others who have made tremendous contributions to society. The goal is to have inventors be seen as heroes in society, just as sports' players and media personalities are. Lee had her team create inventor trading cards, which are like baseball trading cards. Incredible inventors are profiled, including a number of women, and these cards have gone to many schools around the country. In addition, Lee is proud of the National Inventors Hall of Fame Museum, located at the USPTO's Alexandria, Virginia, headquarters. The Museum tells the story of these amazing inventors, the businesses and industries they built, and the critical role that intellectual property played in their innovation. Her office also hosts the finals of the National Inventors Hall of Fame Collegiate Inventor's Competition, where college students from all over the country compete for prize money and recognition based upon their inventions. Their submissions are evaluated by experts from industry, academia, and the USPTO, as well as National Inventors Hall of Fame inductees.

Retaining Top Technical Talent

Lee began her career working as a computer programmer for the MIT Computer Science and Artificial Intelligence Laboratory and Hewlett-Packard's

9.3 Building the Foundation for Innovation 243

Research Laboratories, and later went on to receive a law degree (JD) from Stanford Law School. Prior to joining the government, she was Deputy General Counsel for Google and the company's first Head of Patents and Patent Strategy, where she was responsible for formulating and implementing its worldwide patent strategy, including building its patent portfolio from a small handful of patents to over 10,500 assets in eight years. As she leads the USPTO, one of the largest intellectual-property offices in the world, with over 12,000 employees and an annual budget of over three billion dollars, she and her team are committed to recruiting and retaining both a talented and a diverse workforce.

While the Center for American Progress has found that women represent less than 15% of executive positions in United States companies, more than 40% of executive officer positions at the USPTO are held by women.[2,3] Lee notes that the government is committed over the long term to the recruitment and retention of top talent. The USPTO does a lot to invest in career development, and it's not uncommon to have employees who have worked there for 30 years or more. Lee notes that, in the private sector, by comparison, it is rare to see individuals, especially in the technology sector, who have been with any employer for 10, much less 20, years. The government has a very long-term perspective about employees and investing in their development even though it may not be directly relevant to their immediate task at hand. Lee notes that the USPTO has a rotation program that allows employees to work in different departments. Such programs expose them to different areas of the USPTO's business, and that in turn ultimately makes them more valuable to the institution and helps to retain top talent.

Lee adds that putting efforts into recruiting a diverse workforce is key to achieving optimal business goals. "When I make a decision, I don't want a room full of people who think exactly how I think. I don't come to my best decisions that way. I want diverse perspectives and experiences, and a thorough discussion where differing viewpoints are encouraged and respected. This enables optimal decision-making." While government agencies invest a lot of time in recruiting candidates from diverse backgrounds, including military veterans, Lee notes that it's an issue the private sector is working on as well.

"The question is, how do you retain your top talent?" Lee believes that the long-term perspective and investment in the USPTO's employees help the agency succeed on this front. In recruiting, the agency thinks of having employees for over 30 years and ensures they grow, are challenged, have a sound work-life balance, and remain continually engaged, all of which helps with retention of top talent.

In a piece of advice addressed to young women, Lee says that studying engineering provides a wonderful foundation no matter what career they pursue after graduation. While it's very rigorous, the discipline and analytical thinking skills one learns in the process will stay with the graduates no matter what they do in the future, says Lee. Moreover, she notes that the projected job growth and salaries in the areas of information technology, science, engineering, math, and computer science are high. "If one wants to participate fully and directly in the, by all accounts, very bright information-based economy of our future, having the skill sets in science, technology, engineering, math, and computer science is certainly helpful," she says. The number of filings her office is getting is tremendous, and she doesn't see the number of inventions slowing down. "Wouldn't we want all Americans—and, in fact, all citizens of the globe—to be part of this very exciting future? Our economy and society will be better as a result."

9.4 Career Advancement through Staying at the Forefront of Technology

"If the company is transforming and you happen to be a subject-matter expert on new technologies, there are big opportunities to showcase your strengths and lead."

Jean McManus is the Executive Director of Product Architecture at Verizon Labs, where she oversees product innovation for emerging technologies.

What follows is an edited transcript of an interview conducted by Rahilla Zafar.

When I started my undergraduate studies, I knew I wanted to be an electrical engineer. My father and brother are electrical engineers, so I had personal examples in my life. After seeing the type of work my dad and brother were doing, I felt I could find a niche that could keep me interested and growing in this field.

As an undergrad, I did not always feel alone as a woman, particularly on the electrical-engineering side, since there were typically a few women in the class. However, in some of the advanced core science and math courses outside engineering, I was the only woman—or one of only two—in the class.

9.4 Career Advancement through Staying at the Forefront of Technology

With the female friends I had in engineering, we all connected and created our own strong support system. Overall, I found most professors to be extremely supportive, and my undergrad experience was very positive.

After I graduated, I joined Contel Federal Systems, which sparked my interest in networking as a discipline. On my first project, I got hands-on experience in a lab, working with prototype networking equipment. I felt focused, and because I wanted to continue to grow in this area, I stayed with the company for five years before deciding to pursue a master's degree. Having this lab experience at work really kept me interested. I said, "There's a lot here, I'm really enjoying this and I want to take it further." This led me to grad school.

My original plan was to go back to school for one year to get a master's degree focused on network and security architecture. It was a challenge getting back into the swing of academia, as I had to re-learn a lot of the math. Despite that challenge, a professor approached me and asked if I had any interest in staying for a Ph.D.

I hadn't considered a Ph.D., but I decided to go for it. Subsequently, I received the Patricia Roberts Harris Fellowship that funded women and underrepresented minorities working toward Ph.D.'s. Through this fellowship I worked as a research assistant, which helped me move along some of my Ph.D. work. I would find an area within the networking side that was interesting and then enjoy drilling into the details. This Ph.D. work gave me that depth.

From a research perspective, I had applied networking experience and associated subject-matter expertise. My professor was focused on the mathematics of queueing and loss networks so it was a mutual win. He was ready to expand his research area, and I brought that applied knowledge. Together we found a topic dealing with the transport of video across networks and the trade-offs of delay, bandwidth, and buffer that had enough depth to get me to my Ph.D.

I found my dissertation to be the hardest part. It was challenging to find something I could move ahead with far enough, to show it was unique or novel. But once I got started, the ideas kept coming. I befriended other women who were studying engineering. We compared notes, and if one of us got stuck, we would go over those challenges together. Because of this network—which also included men—I never felt isolated during my Ph.D. work. I think it's important to have that support system to bounce ideas off people and discover new ways to look at things.

Lessons for Growing as a Leader

As I noted, my dissertation focused on delivering video over networks. In the mid- to late '90s, Bell Atlantic offered me a position. Because the company was working on a video-network deployment in Toms River, NJ, it was an exciting—and well-timed—opportunity.

At Bell Atlantic, I worked with senior engineers and was exposed to new protocols and new aspects of designing networks. I eventually moved to a DSL network project and became the lead architect. By then I had set my sights on becoming a network architect, so my work supported my career goals. And I learned a ton! At that time, DSL and its associated data network were the first foray into broadband and data, and many people in the company did not have the experience working with it that I had. There was a lot of room for me to expand my role and grow.

As a result of that experience, I learned an important lesson: If the company is transforming, and you happen to be a subject-matter expert on new technologies, there are big opportunities to *showcase your strengths and lead.*

Look at the technology shifts during the past 20 years. They show us that you grow a lot faster if you don't stay in areas where technology is becoming obsolete. You need to challenge yourself to work with new technologies.

When I first started my career in data networking, I was still trying to learn a lot about scaling networks, equipment, routers and switches. I found that, while you come in with a certain perspective, as you see the way things move and what's coming down the road, your perspective shifts. I was network-focused early on in my career, and I've seen a shift to software defined networking (SDN) applications and open source. All of that is a huge change in the way I think about my career, as well as about the innovative solutions and services we offer our customers.

I always tell younger developers and engineers to look for those technology shifts and read ahead. Five years or more ago, if you read up on SDN, tracked the forums digging into it, and followed the different approaches, you would be ahead of the curve because now providers are rolling it out.

Another challenge is *knowing the right time to switch your career focus.* When I first transitioned into management, I had a team of architects focused on both carrier Ethernet and the IP edge. Verizon was evolving, so it was a great opportunity. But when you're a subject-matter expert, no one wants you to leave and create a gap, so it can be very hard to step away. My challenge was getting stuck doing something for longer than just a few years. It's important to think carefully about what's next for you when you're ready for a change. You want to make sure you transition into something that mirrors your career goals. If you want to stay technical as a network architect, then opportunities in program management or operations may not be the right fit. If you've been working on IP-layer networking, for example, you might consider moving to the optical networking group and learning a new layer or set of technologies.

Women Advancing in Engineering

Engineers often want to give everyone working with them credit. This tendency is a challenge for women. The minute a woman says, "so-and-so helped," others assume those people—not the woman—did the real work. When you cover a technical topic, you need to make sure you put your name on the presentation. If someone comes up to you and congratulates you on a good job, don't list all the other people who worked on the project. Just thank them and talk about where you see the project going. Keep the focus on your contributions. Most men would not list everyone who worked on the project with them. This is the difference between males and females: women are more inclusive, and men are not afraid to say, "Hey, I did this."

We need more women in the electrical-engineering field. As a technical woman, you can feel isolated at work. In my career, I've enjoyed a lot of good support from men who have connected me with people and helped me get on the right projects. But a lot of work gets done in informal conversations, or over beers after work. Women don't often get invited, or if they do, it's a different type of conversation. As more women enter this field, our network will become stronger, and it will become easier for women to find opportunities that let them grow.

At my first job, I was surrounded by many senior technical women. There were at least 10 of them. Many were pursuing master's degrees in computer science or electrical engineering at night. In that setting, it wasn't strange that I was a female engineer. Those were the people who kept me going on the engineering side and became my role models early in my career.

I found that there were more women on the operational side than in architecture. Even then, we felt we needed to bring more women in, and that trend continues. Today's shift from a technology focus to software has opened the door for women who have the ability to code and work as developers.

How do we retain technical women, give them opportunities, growth paths and the ability to see other technical women who have succeeded? The answer is to plan events that connect women so they can build their own networks, be exposed to female leadership, and learn how to navigate in a larger company.

Allowances for personal time: If you're at a large company, certain projects could have you working really long hours in the lead-up to product launch. Make sure you allow people to experience important times in their lives. If someone has just had a baby, parental leave is important for both women and men. I personally like to see an emphasis on going back to school to pursue advanced degrees or change degrees. Programs that help people seek further education and higher credentials at night are important for companies to support. There will be times when these people need a few hours to study, complete major projects, and so on, if they take advantage of these programs.

Younger workers: Today, most people don't set out to work at a single company for 30 years and then retire. Younger people—and especially millennials—come to a company, gain experience, and then find the next thing that will help them grow. At Verizon, we bring in people who have moved around a lot from a job-experience perspective to gain a broad view of ideas that can help us understand what people might want from a product and offer different perspectives to the technical solution. It's a bit like open-source software, as it allows everyone to contribute and offer their unique expertise.

Diverse teams: Shouldn't we want the people designing our products to represent the people we sell them to? If we don't bring in diverse teams, we miss a big opportunity to see other ways to use the products, features that will be helpful, and so on. This is true on the network side as well, because it can lead to better architecture. Perspective is everything. With diverse teams, you can have robust debates that lead to solutions that can scale, meet customer needs, and be valuable to the company.

Shifting into technical management: It is always a hard decision, whether you're a man or a woman. If you're interested in management, you should conduct a self-assessment on your own skills and interests. You may feel pressure to manage people, but maybe it's not the best fit for you. Are you comfortable operating on a much higher level of detail? Can you go into meetings not knowing every technical fact but still provide a broad view? Do you like to deal with people, help guide their careers, manage assignments and offer candid feedback on performance? If these options don't appeal to you, you may be better off as an individual contributor.

9.5 Being a Woman on an All-Male Executive Board

By Laurie Cantileno

"There's power in relating to individuals who want to take a chance on you. Be open to that."

Sue Quackenbush is Chief Human Resources (HR) Officer at Vonage.

Sue Quackenbush is the Chief HR Officer for Vonage. She leads all aspects of HR, the company's corporate social responsibility, and foundation. She has been working for about 27 years, spending the last 19 in HR after eight in corporate finance. Prior to joining Vonage, she was the Chief Human Resource Officer at Presidio. She's held several other leadership roles throughout her career, and has also worked for Thomson Reuters, in a role that ended up being pivotal in her career.

A Winding Path to Technology Leadership

Sue had an unusual upbringing. She grew up on a very small family farm of less than six acres, 12 miles outside New York City. She was the first child in her family to ever attend and graduate from college. She had not been exposed to the corporate world or the technology field until her sophomore year in college, when she went through the school's cooperative education program and worked full-time for a semester and a summer at the Celanese Corporation. After beginning her career in finance, Sue changed career paths unexpectedly and entered HR, where she spent the bulk of her career, primarily in the technology industry.

It may seem as if Sue's journey to leadership in the technology space took an unusual route, but for women of Sue's generation and mine, a winding path to technology leadership is a very common occurrence. There were not many things afforded to us to aspire to or emulate in these fields "back in the day." Had there been opportunities, I am certain that many women's careers would have progressed differently, and had women had mentors, then sponsorship would have become more common. This is a key point in Sue's story.

"When you look at how women can take advantage of technology," Sue observes, "and at some of the qualities a woman needs in order to be successful in technology, I would say first and foremost, it's being open to change and to the art of the possible. In my view, these are the attributes that any woman, any person, especially in technology, needs to have in order to thrive."

Sue goes on to describe this theory of sponsorship or what having a "champion" really means. She advises to look out for your champion, who is more than a mentor and a coach. It's someone who believes in you and will not only support you, but will take a risk on you to do something that is out of the ordinary. "My first champion came to me at Thomson (now Thomson Reuters), and his name was Bob Daleo. He was the Chief Financial Officer (CFO) at the time, and championed me to make that career switch from Corporate Finance to a Human Resources and Organizational Development role," she explains.

Sue left Thomson as a full-time employee and came back as a consultant after her first daughter was born. She needed to take a step back to care for her daughter. While consulting, Daleo asked her to come back full-time to help him with organizational development work, and that's how she transformed her career and started on the path of becoming a Human Resources leader.

It's important to keep in mind, Sue says, that not every mentor is capable of being, or willing to be, a champion. She also recommends staying positive and not exuding negativity. Someone acting in your best interest or on your behalf wants to have confidence in their choice, so it's important to stay as positive as possible in your dealings with your sponsor.

"There's power in relating to individuals who want to take a chance on you, so be open to that," she adds. Another key to success, she says, is being comfortable with ambiguity. In all of her jobs within HR, she recalls, she's taken newly created roles, which she's then had the opportunity to shape. "When you can operate with ambiguity, you get to really define your role and experience, so you have to be comfortable with doing that," she adds.

Sue has been married for 25 years with two teenage daughters. Her oldest daughter was born with an immature nervous system and colic. This caused her to cry for almost 16 hours a day for the first 10 months of her life. Eventually, her baby's doctor recommended she leave her position to care for her daughter as her health wasn't progressing. "I ended up resigning from my position since I could not commit to returning to work at full capacity. It was a difficult decision for me to make at the time, but really, it was the only decision and, ultimately, it changed the trajectory of my career in a very positive way," she explains.

Sue ended up being out of the workforce for a short time only. Before long, the company was asking her to come back as a full-time employee. Before going back to work full time, Sue needed to figure out how to make it work. She and her husband assessed the situation from a financial perspective. There was more opportunity and earning potential for her husband at the time than there was for her.

This was one of the more profound life learnings for Sue. "When you're in the work world and in a professional career, you work hard and want to keep advancing, but when you get thrown an unexpected life situation, no matter how much you try to control something in your professional life, your non-professional life needs to be in balance. Life happens, and you have to figure it out," she says.

Sue explains, "The beauty of being in the role I'm in today is you do get to regularly reevaluate all of the offerings that we provide to employees. I am looking at what's progressive in the technology space but also at what's balanced and fair. The challenge is always to determine how an organization attracts and retains the right talent. My current role allows me the ability to draw on my past experiences, professional and non-professional, to help build an environment where employees can grow their skill sets and their career, and flourish."

9.6 Gender Bias in the Tech Sector

By Laurie Cantileno

My influence in engineering came from my dad, who was an engineer and who would always say to me, "You can do anything a boy can do and do it better." Then in his next statement he would say jokingly, "I would trade all my three daughters for just one son." He was a very old-fashioned Italian and thought women should only leave home to get married, and his advice to me was two-fold when I told him I wanted to be either a hairstylist or work for NASA. "Do not become a hairstylist because you are on your feet all day and it's a waste of your brain, the space program is dead so learn computers. It's the way of the future and maybe you will find a nice husband at school too." So I decided to try it, even though I had never seen a computer (no laptops in those days). When I enrolled in a computer-science program at the New York Institute of Technology (NYIT), I was one of only three girls, and so began my interesting journey in technology.

Margaret Chiosi, Chief Architect at AT&T, had similar experiences and advice. Her father, too, was an electrical engineer, and her mother was head of a Chinese news bureau in Washington DC. As Margaret tells it, "Chinese parents with college degrees in the United States always focused on STEM. When they moved to the United States, their generation pushed all our generation into STEM," she explains. Mirroring my own background, this is another example of how parents are huge influencers, cheerleaders, and advocates.

There are so many obstacles I ran into—harassment and prejudice—and, looking back, I'm amazed I never gave up. My nature never let me, and in fact it made me stronger. The first day of college, two guys hit on me and almost got into a fight. During the first professional out-of-state training class

I ever went to, my teacher showed up at my hotel room with a gift. I was always the lone woman or one of very few. My long and well-styled blonde hair was pretty apparent when I walked into the room (remember, I had wanted to be a hairstylist), and I recall one time during another training class, I walked into a room full of men and the eyes went from top to bottom assessing me even before I opened my mouth while quietly dismissing my technical credibility.

As my career progressed, it just became worse. There are so many stories, and these situations shaped me into a confident and somewhat aggressive woman. It also shaped my wardrobe. I started to dress more and more like a man and played down my womanhood altogether, so as not to distract anyone so they would see me and not my female stature. It was an experience that Chiosi went through as well: "A woman Assistant Vice President (AVP)—this is when women were just starting to move up the technical ladder—told me once to stop smiling because I would not be taken seriously. She was trying to get women to act more like men."

Slowly I made my way up the executive chain, weaving in and out of those types of incidents, until one day I landed a sweet role at a financial services company as VP of Engineering, Operations, Collaboration, and Security, reporting to the Chief Technology Officer (CTO). He was a raging hothead, cursing at every moment when something didn't go his way. He continually reminded me I was the highest paid of his all-male staff, and I knew it was killing him. But I needed to be tougher than that and be "one of the boys." Until one day, he and another raging Senior Vice President (SVP) slipped up and I was not going to take the harassment any longer. This was a key turning point in my career (Sandberg's book being the second).

The SVP of one of the business units my teams supported ran into me with my Senior Director (Chris), who reported to me, in the cafeteria. There was a gentleman on my staff he just didn't like and whom he thought was completely incompetent. So he told me to fire him without warning, just like that on the spot. Well you just don't do that, it goes against all human-resources policies. This person in no way reported to him and his claim was unjust. But this SVP was known for his irate and erratic actions and no one ever stood up to him. He approached me and then the conversation turned quickly as he screamed and cursed at me, getting closer and closer while I retained my composure until Chris stepped in front of him and said he needed to stop. Everyone in the cafeteria was horrified and I was angry and shaken at the same time. I didn't think he would've behaved this way if I were a guy because he knew no man would have put up with this behavior. It was a cruel way of bullying.

It only got worse. This SVP proceeded to tell my CTO that I was incompetent for not firing this individual. I told him about the incident and reported the SVP for his horrid behavior in front of a cafeteria full of people. Even though I had witnesses there to affirm what had happened, he proceeded to demote me, stripping me of half the department staff I managed without warning. Needless to say, this was illegal. I resigned and was given a large sum of voluntary compensation, which to me was a clear sign of admittance of wrongdoing by the company.

The company did offer to transfer me to another department but I declined, explaining that I shouldn't be the one having to leave, it should be the people who harassed me. This was the very same company who selected me to be on the Board for Inclusion and Diversity. After my incident, it didn't take long before many women came forward and told me similar stories that they had been too afraid to speak out about. I learned from my remaining colleagues and several other female employees who came to me privately after I resigned, that the company has had several gender lawsuits against them but squashed every one of them. Soon afterwards, the SVP in question was investigated by the president of the company and encouraged to retire, because there was another incident where he proceeded to call a woman the "C" word.

Yes, this really does exist in corporate America to this day, even with all the "awareness" we have. Sexual harassment, sexual innuendos, and corporate bullying are alive and well. Oh and by the way, another female VP quit after me as a result of blatant gender bullying, even though she had the highest sales in the company, was a graduate of the Wharton Business School, and has been seen as a rising star in the company. In the meantime, I had started my own management-consulting firm and bought a Porsche. Personally I think both those men did me a huge favor, because if it hadn't been for that incident, I would probably not be writing this chapter today.

9.7 A Snapshot of Women in Chinese History

By Aglaia Kong

Aglaia Kong is CTO of Corporate Networks at Google. Prior to this role, she was CTO for IoE (Internet of Everything) Solutions and VP of Technology, leading solution innovations for IOE, at Cisco Systems.

"The big leap for women's leadership has become quite pronounced over the last 35 years, as China has gone through explosive economic growth."

China's transformation has not been easy. Historically, there were short periods in which a few women became national leaders, such as during the Tang Dynasty (618 to 907), and the Qing Dynasty (1644 to 1912).

Throughout Chinese history, many rules have been imposed on women to keep them in check. Some have been extreme, such as foot binding, a ban on education, a prohibition on women to leave the house once they have reached marital age (which was once 12 or 13), living with their husbands when they passed away, having to live in polygamous marriages, and being untouchable by physicians during check-ups, thus making medical treatment challenging.

The People's Republic of China (PRC) came into being in 1949. Before that, from 1912 to 1949, the Republic of China included mainland China, and sometimes, Mongolia and Taiwan. Up to 1912, China was ruled by the Qing Dynasty and suffered many years of civil unrest till the Dynasty came to an end with the Wuchang Uprising and the Xinhai Revolution in 1911.

Overall, from 1800 to 1949, many subtle changes helped set the stage for women's empowerment in China. Qing's origin was Manchu, not Han Chinese, which did not come from a culture of suppressing women's abilities. The last few Emperors in China were controlled by Empress Dowager Cixi, who ran the dynasty behind the scenes by carefully selecting emperors who were young and could be controlled. Lastly, in the late 1800s and early 1900s, China began participating in many international exchanges, thus becoming more exposed to Western cultures.

When the PRC was finally established in 1949, Chairman Mao faced many challenges, including a low population because of years of war, not enough adult labor, and the departure of many educated professionals from mainland China to live in Taiwan, Hong Kong, and elsewhere. Mao set policies that paved the way for the culture of gender equality that is seen in China today. In addition to emphasizing how women should hold up half the sky, thus allowing them access to the same jobs and the same education as men, he made the first 12 years of education free. He also called for a re-design of Chinese characters to make learning to write easier.

Those policies worked over time until the Cultural Revolution, which took place from 1966 to 1976, and put China's education transformation to a complete stop. The whole country was focused on class equality. However, many people used this movement to remove people with opposing opinions and who dared to question authority. During this period, many educated professionals were punished, killed, or sent to labor camps. There was no proper education from elementary school to college. People had limited access to information and no access to the outside world. Only a handful of books could be read and only a few movies could be watched.

A Leap for Women's Leadership

China started to open up to the outside world after Mao passed away in 1976. In 1979 Deng Xiaoping started the economic reform once he came to power. The big leap for women's leadership has been quite pronounced over the last 35 years, as China has gone through explosive economic growth.

Starting in late 1978, China was opening up by providing a platform to drive economic reform. Policy changes enabled many people to start opening small businesses. Women owned small shops selling food and clothing, as well as different types of service- and entertainment-related businesses, which also became quite common.

These small businesses helped many women make a strong start. Women also got into real estate: buying and selling properties, investing in malls, and building bigger factories. A few women did very well out of these new opportunities, but most still lived in villages and inland cities of China, and were left out of the economic growth that was taking place mainly in large cities. That expansion did create a lot of jobs, though, so many women from the poorer parts of the country would migrate to work in the bigger cities. Many of these jobs were, again, service- and entertainment-related. The money these women earned helped their families back home. Many women also used their earnings to start small businesses back home as well, and became better off over time.

During the same period, many international companies started to move their manufacturing facilities to China, in part because of the poor regulatory environment (there was a lack of environmental-protection laws, for instance), and vast supply of cheap labor. As a result, many jobs were created, thus leading to the creation of opportunities for women. The period from 1980 to 2000 saw dynamic growth, and it seemed that everyone's main goal was to get rich by any means. But soon people recognized that, to sustain the economic growth, proper education, and the use of technology to leapfrog over barriers to development, would be needed over the long term.

This period saw the rise of technology companies in China, but it was challenging to jump start since most companies were focused on manufacturing and provided low-skill jobs.

During the same period, many international companies were confronting the challenges of not being able to hire enough skilled workers to satisfy the needs of the technology boom. And as international companies discovered that China produced three times more engineers than the United States each year, waves of out-sourcing began.

With a shortage in the hi-tech labor force, many international companies started setting up research-and-development centers in China. These investments helped train Chinese technical leaders and paved the ways for China's technology boom. In turn, the Chinese government began to get very conscious about local innovation and intellectual property owenership. A number of good practices emerged from this shift in policy, including the following:

- Every five years, a national five-year plan sets out areas of focus and sets aside funding for each of these. For example, in the current plan the focus is on education, healthcare, and environmental protection. Each area gets from 3 to 5% of the GDP as an investment fund to be used by companies for innovations in those areas.
- The government has also created many incentive plans to encourage investment in new technology. For example, green energy is now being promoted as a way to reduce pollution and encouraging green energy technologies is an important goal. Examples of government incentives are free land and tax incentives for anyone who wants to build solar companies, thus helping make China the number one producer of solar panels in less than three years.
- Strong policies have also helped build giant companies such as Huawei, Tencent, and Alibaba. These companies all have very diverse leadership. [Huawei's veto-wielding CEO is Ren Zhenfei, and it has three rotating CEOs: Mr. Hu Houkun, October 1, 2014, – March 31, 2015; Xu Zhijun, April 1, 2015, – September 30, 2015; and Guo Ping, October 1, 2015, – March 31, 2016. All are men.] Almost 34% of Alibaba's high-level managers, a third of its founding partners, and more than 40% of its total workforce, are women.[4]

I was born and raised in a small village in the southern part of China during the Cultural Revolution. Because much of the educated professional class was afraid of persecution, my school had only one teacher for grades one to six. She had to teach math, science, Chinese, history, and geography—even though she had no training in most of these subjects. She also needed to grow her own food, and we could not help her since she was deemed to be from a lesser social class as a teacher. Needless to say, we did not get much education from school.

I was both lucky and unlucky during my childhood. My parents were college graduates. My mom was a chemistry professor, and my dad was a mechanical engineer in charge of designing and manufacturing trackers. They were treated badly during that period: sent to different parts of China to work, and could see each other only once a year. My sister, my brother, and I were not allowed to go with them so we were raised by my grandmother in a remote village and were visited by our parents only once a year. So I was raised as a wild kid with no boundaries in creativity.

Time in the village was fun but scary. My grandma was classified as a land-owner and that was considered to be bad. So we lived in constant fear that she might be taken away and we would have no one to take care of us. But at the same time, I was trained to be brave and strong to protect her and my younger siblings.

My dad was able to introduce me to the world of science. He hid his college books (all in Russian) in a secret location at home. I went through the books on my own. Despite the fact that I had no clue what those books were about, they still provided me a glimpse of something beyond what my school was teaching us. After my dad found out that I was reading through them, he started secretly collecting unauthorized books for me. I would wait eagerly for him to come home each year so I could read them. Those books opened a world of wonder for me. One thing that really resonated with me from reading them was that we might not be alone in the universe, there were others out there too.

I started looking up at the sky, thinking about extra-terrestrials. I was hoping to get "picked up" by alien ships. Waiting on the roof for seven days while getting bug bites everywhere, I decided to take matters into my own hands and make my own spaceship to go to outer space! I asked my dad what I had to do to make my own ship; he said that I at least should be an electrical engineer. So at the age of eight, I set out to become one. That goal motivated me throughout my entire life. I feel privileged that I got the opportunity to pursue my dream. Around 1979, my family was able to leave China and moved to Macau. I went to Canada for high school and later pursued my university education in the United States.

I eventually got my electrical-engineering degree and started doing advanced research in chemical engineering with lasers. Laser research helps NASA to use the right type of material to build its space shuttle. And the same research is used to guide the shuttle to dock with the International Space Station. Beyond that, I am also a pioneer in the use of GPS for navigation. I have worked on storage- and file-system-virtualization technologies, cloud computing, and networking. I also worked as IOT CTO for Cisco Systems, defining solutions for next-generation cars, planes, trains, and cities by connecting everything together. One of these days, I will make my own spaceship to explore what is outside of Earth. When that day comes, I will look back with my vivid memory of my childhood days in China.

References

[1] http://www.uspto.gov/about-us/news-updates/remarks-director-michelle-k-lee-million-women-mentors-summit-and-gala
[2] https://www.americanprogress.org/issues/women/report/2015/08/04/118743/the-womens-leadership-gap/
[3] http://www.fastcompany.com/3047424/how-former-googler-michelle-lee-plans-to-supercharge-the-us-patent-office
[4] http://fortune.com/2015/05/20/alibaba-jack-ma-women/

PART VI

Emerging Fields of Technology

Introduction

A whole new wave of emerging technologies is not only going to change how we live but could potentially foster a whole new set of female experts. In this section we highlight women who, by challenging traditional funding models, are creating a level playing field in new technologies, from artificial intelligence to blockchain and environmental innovation. We hear from some of the world's most respected experts and practitioners on the opportunities for women in the field of cyber security and cyber psychology. These leaders also highlight solutions for engaging girls and women, ensuring they have a pathway to enter into these exciting fields regardless of what stage they're at in their lives or careers.

10

Defining the Cutting Edge

10.1 Meet the Woman Who's Created the 21st-Century Finance Model for Emerging Technologies

Riva-Melissa Tez is the CEO and co-founder of Permutation in San Francisco. A London native, she runs an artificial-intelligence platform and incubator. In her spare time, she works with Maria Konovalenko and Steve Aoki on The Longevity Cookbook, which distills academic research into practical measures for slowing aging.

What follows is an edited transcript of an interview conducted by Rahilla Zafar.

I learned important lessons about money at a very early age.

At 10, I moved into a homeless shelter after my father left my mother. My mother is severely schizophrenic—which can be both chaotically fun and devastatingly traumatic—and was not well enough to look after herself, let alone look after me at the time. Amongst other things, she used to make me drink the milk in the morning first to check if it had poison in it. A few years later, at 14, we moved into social housing. We lived in this horrible apartment that had no curtains or carpets. At one point I realized, "Oh, money is how the world works." It's a lesson you don't learn at school.

I received a scholarship to attend a prestigious all-girls school. The girls at my school came from quite wealthy families, so I never told anyone where I lived. I would take different routes walking home so that no one knew where I was going.

Once I realized that money was the key to an escape, I started reading books on consumer psychology. Edward Bernays, who was the nephew of Sigmund Freud, was a great inspiration to me. At 15, I managed to get a job selling outdoor sinks at a furniture trade show. The job initially belonged to my older sister, but on her first day of work she fell asleep on the job and was promptly fired. The company told her to find a replacement, so I begged her to let me do it.

My sister lied and said I was 16. I ended up selling these sinks at the show from 9 a.m. to 9 p.m. for three weeks straight. I developed my own strategy to improve my learning: I would read books on sales and then go test theories out when I was working. Later, I would update my model of consumer psychology, which I kept track of in a notebook. In this way, I used very rapid AB testing to learn about sales.

I sold more at that trade show than the entire management team, earning £15,000 in commission, which was more money than I could even fathom at the time, considering that my mother and I lived on £60 a week. I had truanted school to accommodate the job and was in a lot of trouble, but my grades were good, and to be honest, I didn't really care much about school at the time. I saved the money and was able to eventually move my mother out of our awful apartment into a beautiful one-bedroom place in West London, where she still lives today. I have paid her rent every month since then. I owe her for a lot of my thinking, so I could never abandon her. Those struggles she put me through gave me a high tolerance for risk and stress, which I am extremely grateful for.

People sometimes ask what made me ambitious at a young age. The motivation was simple: we were literally so poor that I *had* to find a way to make money for my mother. There's no secret sauce; it was just pure desperation. There were times when we couldn't even afford to eat.

I attended University College London and switched from a joint Honors to straight Philosophy, with a focus on Epistemology and Logic. My mother's condition got me hooked on trying to understand how humans justify their beliefs.

Whilst at University, I was living above an empty shop in Notting Hill, which is an affluent area in London. During my time there, I noticed there were many families living in the neighborhood. One day, I suggested to my boyfriend at the time that we should open a pop-up toy store in the empty store below. We followed through, and surprisingly we did very well—so much so that we launched a permanent store. Eventually our shop expanded, taking over the two stories. Our toy store still stands today. It has just celebrated its sixth birthday.

At 21, I got more serious—I thought, "Wait, but I don't want to run a toy store for the rest of my life . . . " I wanted to learn how to do things digitally. I used to get into trouble for hacking stuff off computers. I was very much

10.1 Meet the Woman Who's Created the 21st-Century Finance Model 265

the kind of teenager who would stay up all night on a computer. This was all inspired by my father, who is a professor of electrical engineering. Before he left my mother and me, he used to make me build computers, and I watched him code when I was very young.

After I left the toyshop, I moved to Berlin with a mission to further my software skills. After a few months into this endeavor, I had the idea of building a creative social network for kids. I worked on it with an American co-founder and ended up staying in Germany for over two years. At one point, I read Kevin Kelly's *What Technology Wants* and got really interested in emerging technologies. I wanted to find people to discuss ideas with, so I started a nonprofit called Berlin Singularity, which hosted talks and events. I also started teaching as a lecturer at two business schools, mainly on consumer psychology but also on entrepreneurship. Some friends and I entered and won the LinkedIn Hackathon, which is still one of my favorite memories from Berlin.

Michael Vassar, who at the time was running the Singularity Institute for Artificial Intelligence in San Francisco, came to give a talk for Berlin Singularity. Afterwards, I complained to him that I had this sense of imposter syndrome. What I meant was this: the thought that people were regarding me as some sort of European expert on emerging technology horrified me. I remember seeing myself in this magazine listed as one of the top influential people in technology in Germany, and I kept thinking, "But I'm only 21 and don't know anything!" I remember Michael telling me that I would enjoy the intellectual climate in San Francisco.

A few months later, I was in a near-death car accident on the Autobahn. I got admitted into the hospital with a bad concussion and kept thinking about how much I wanted to learn about the world. A few days later, I left my house, start-up, and boyfriend behind in Berlin and flew to San Francisco.

An interest furthered by the near-death experience, my original plan was to study how I could contribute towards life extension and regenerative medicine. To begin, I started profiling all of the biotech companies I found interesting. Trying to fulfill the challenge I set myself in the hospital to learn about the world, I also picked up a hobby of reading a lot of physics papers. To this day, I find them very humbling.

Eventually I began to work with investors who wanted to put their money in emerging biotech opportunities. As I started to undertake technical due diligence on different companies, I realized that solving scientific challenges depended on the strength of the tools and resources that researchers have at their disposal. It was then that I really hit on machine learning and artificial intelligence (AI). I kept thinking, well, if you could improve AI resources, you could use it to solve complex problems, such as those in biotech and energy, among other industries. AI has the potential to make the scientific process cheaper and *better*. It has the power to reduce complexity, which is an idea I think about all the time.

Once I'd hit on AI, I knew what I had to work on. I believe AI is fundamental to improving humanity's odds of survival because humans are limited in their scope of skills for problem-solving. The only other option would be intelligence amplification *en masse*.

Women in Science

The subjects that we work on tend to fall under the STEM (Science, Technology, Engineering and Math) categories that already have poor gender rates. I think the disparity is improving today, because there are more opportunities to learn about engineering for children.

Today, there are so many children's toys that promote engineering for girls. This is in sharp contrast to 10 or 20 years ago. Today, we're at a point where we are just ahead of the curve in getting women into STEM subjects. I believe the ratios will balance—it's just a matter of time.

Frequently, I'm the only woman speaker at an event or on a panel. I recently shot an ad campaign for Microsoft where I was the only female. People ask me what that's like all the time, but most of the time I don't even notice. Sometimes I get talked down to, but I just play into it. At this one high-profile investment event, guys kept asking me where my husband was. I guess it was impossible for them to believe a 26-year-old woman could be an invited guest. I told them that he had stepped outside to make an important business call and made a note to never work with those people. I don't feel a need to justify myself.

Creating a 21st-Century Finance Model

When I first came to the Bay Area, I was working with two kinds of groups: investors who wanted to put their money in biotech, and start-ups that wanted help raising money for their projects. As a philosophy graduate, I didn't know a lot about fundraising.

I spent a great deal of my time learning about how venture-capital funds work. The traditional fundraising model didn't make much sense to me. I saw it this way: as a venture capitalist, you're basically a kind of middleman who invests in start-ups. You represent other people's money, which means you have to answer to their particular interests. But, at the same time, you also need to negotiate the interest of start-ups and their entrepreneurs. I was pretty open about telling people in asset management that I didn't entirely understand some of their models, and that some of them might be more imperfect than others. Then, someone rightly challenged me on how I would do it differently. I told him I would learn everything I could about venture capital and attempt to create a new model.

I left America and spent nearly three months in France reading about venture capital, including all of the venture-capital reports from the Kauffman Foundation over the last few decades. At first, I didn't even know what a Limited Partner was. When I had finished, I could break down the Internal Rates of Return (IRRs) of the top venture-capital funds over the last 10 years. I just studied everything I could and concluded that the model of traditional venture capital didn't work for the kind of projects that interested me. It works well for consumer tech, but it doesn't work as well if you want to do anything long term.

I wrote a few posts about how we should reshape finance. They got over 250,000 views. As a result, a couple of big investor groups were interested in hiring me. I found that ironic, but then again everyone loves a contrarian. Among those who reached out to me was Peter Bruce-Clark, who was working at Stanford looking at innovative investment models. I ended up spending much of my time talking to him. He walked me through some papers, including research on new models for investment funds that he himself had been building, which were very impact-focused. I decided I wanted to do the same thing for AI and began working with two others on fleshing out that model, which was a hybrid impact investment fund.

My original goal was to raise these funds to invest specifically in artificial intelligence. This was about 18 months ago. During this time, a smart investor and mentor said to me, "The bottleneck in solving problems such as the advancement of AI isn't a lack of capital. The problem is a lack of valuable deal flow." Essentially, he told me that I wasn't tackling the real problem. It was a major shape-shift for me.

I decided to take another six months to model the AI ecosystem. I wanted to learn about where talent goes, the flows of capital between companies and investors, why certain companies are built, and why some ideas don't get off the ground.

One of the problems is that investors don't understand much about research-driven machine intelligence since it's extremely technical. I saw a market gap for better due-diligence tools as well as a gap in the provision of services geared to assist the early-stage development of AI companies. It's a much better, more valuable use of my time to ensure we have lots of smart AI investors rather than trying to launch a stand-alone fund.

That's how Permutation ended up being developed. I created a suite of tools, which we're launching soon, and which help investors with some tricky decisions. I'm also trying to introduce investors to more hardware-based projects. People focus on top-layer innovation, such as algorithms, without recognizing that the very thing that sets the parameters for improvement is the hardware. I'm lucky that I can call up my father when I need some help understanding a quantum computing paper.

The second thing we're doing is investing in and incubating early-stage AI companies. It's a very compelling message when you go back to investors and can demonstrate how you've built tools to measure your returns and your ecosystem impact. Typically, you raise a fund and *then* you try to do that (or perhaps at least pretend you care about impact). We did all of the research first and remain extremely research-driven in our investment model and thesis. Nobody we've approached for funding thus far has said no.

Peter, who helped me learn about measuring impact returns, left Stanford and is now my business partner and closest friend.

There's a lot of hype right now around AI, so while there are many interested investors, we're seeking smart ones that pass different integrity tests. The same holds for the people we hire. I'm looking for impact-driven individuals and groups who genuinely care about making the world better. Making money isn't as hard as creating a positive global impact, although capital can certainly fuel the strength of impact. Unfortunately, I think most people realize this in life when it's far too late and they wish they had done more with their time-slice of existence. I'll retire when I die—and I'm not really planning on doing that either.

Incentives and Collaboration

Last year, I went through a bout of depression because it bothered me to see how little people in power didn't collaborate to solve some of our grandest challenges. There's too much ego and too many opposing forces, not to mention incentive structures that work against us. I kept thinking: how can you make collaboration easier without losing the value you have as a company or the impact you have as an individual?

We haven't quite worked out how the human mind works well enough to understand how to manage incentives. The problem is, we incentivize people *not* to collaborate with others. You collaborate within your own group or within your company, but if someone does something similar to you, you want to deliberately withhold information from them. It restricts everyone and restricts innovation. This kills research progress.

10.2 Women Leading the World of Blockchain

Rahilla Zafar co-authored this piece with Meltem Demirors, who is a Director at Digital Currency Group (DCG), a company that builds and supports bitcoin and blockchain companies. DCG has invested in BitPesa and BitOasis, all highlighted in this piece. Ameen Soleimani and Martin Lundfall are co-authors as well and both Software Engineers at ConsenSys.

One of the most exciting innovations in recent history has been the emergence of distributed computing systems, and in particular, digital currencies and the underlying technology known as the blockchain. Since it launched in 2009, Bitcoin, the first and most prominent blockchain-based digital currency, has demonstrated the viability of using the blockchain—a globally shared transaction history along with a set of rules that allows for anyone to add new, valid transactions—in order to manage asset transfers in a secure and transparent way. Originally, the Bitcoin blockchain was only used to track bitcoin transactions, but entrepreneurs have begun to use blockchain technology for other assets like money, stock bonds, loyalty points, music files, a deed, a license, or anything that has value. Anytime an asset changes hands—whether it be a financial asset or a digital asset—that change in ownership is recorded on the blockchain, and witnessed by each computer or node in the network. The striking feature of this process is the way in which participating parties come to agree on the current state of the network. Whereas transactions in the traditional financial world always need to be facilitated by an intermediate institution like a bank, blockchains enable their participants to self-govern, making it expensive for any singular entity to control the network. The blockchain is extremely difficult to corrupt or censor; tampering with historical records (perhaps to falsely claim ownership over an asset) requires computational resources greater than the combined resources of the rest of the network, which for Bitcoin at the time of this writing would cost roughly \$150M.[1]

Today, 50% of women around the world do not have access to bank accounts or financial services.[2] For the first time in history, it is becoming economically feasible to extend financial services to a very long tail of consumers who may not be profitable for traditional financial institutions.

Digital currencies and blockchain technology, coupled with the proliferation and spread of smart phones, are making it possible to provide banking services, credit products, investment opportunities, and a whole suite of solutions to women and other underserved groups across the world either through business-to-business, business-to-consumer, or peer-to-peer (B2B, B2C, or P2P).

Many start-ups have taken matters into their own hands, leveraging the new opportunities of blockchain technology and the low-cost, low-friction digital currency payment rail. Now a woman in a village that is hundreds or even thousands of miles from the nearest bank can receive digital currencies such as bitcoin, or digitized government fiat via her phone, save money, apply for microloans, make investments, and pay for goods and services, thus enabling her and others to become financial decision-makers, and financially independent.

The innovations underlying blockchain have enabled a wave of innovation beginning with money and the rise of digital currencies and has spread to data storage, digital-rights management, identity, and dozens of other assets and data types. Digital currencies represent an $8 billion financial system[3] that has evolved without banks or financial institutions—just a community of software developers building open source code and early adopters using this "digital currency."

Blockchain technology is demonstrating its ability to help facilitate the movement of ideas, digital goods, data, content, and money in simpler, safer, and more-transparent ways. According to a report from Santander, a global bank, blockchain technologies could reduce banks' infrastructure costs by $15 to $20 billion a year by 2022,[4] mainly because of the reduction of back-office expenses that this new technology makes possible. Digital currencies and blockchain technology could also create new revenue opportunities by enabling access to financial services at scale, through mobile phones for today's two billion unbanked working-age adults.

Beyond just financial assets, nearly one-fifth of the world's population does not have a formal government identity, meaning nearly 1.5 billion people are,[5] in that sense, invisible. The United Nations recently created a special taskforce, ID2020, focused on providing legal identity to all by 2020, which includes blockchain entrepreneurs, policymakers, and NGOs. These groups are exploring the use of innovative identity technologies that protect vulnerable populations around the world and offer them access to legal and political protections. At the ID2020 Summit held at the UN Headquarters in New York City in May 2016, Microsoft, Blockstack Labs, and ConsenSys, announced a new collaboration to create an open-source, self-sovereign, blockchain-based identity system.

Christian Lundkvist, lead software architect of ConsenSys's identity platform, uPort, noted: "I'm very excited about this collaboration, which promises to radically expand the reach and user base of self-sovereign digital identity systems. With this project we are taking a big step towards empowering people who suffer due to the lack of identity, as well as streamlining the fragmented identity systems in our modern society."

Carolyn Reckhow, director of operations at ConsenSys adds, "Outside of the human rights implications that sovereign identity brings for disenfranchised populations, what really excites me about uPort are the new business models around putting personal data back into the hands of the people. In a world where AI systems profit from our personal information, blockchain identity solutions present an alternative vision of the future where people can be enriched from their data through privacy and granular permission systems, rather than threatened by the increasingly centralized data silos in the hands of major tech companies."

The adoption of blockchain technology and its increasing number of applications, from digitized asset transfers to the backbone of the financial system to identity management, has been greatly accelerated by advances in blockchain technology itself. In 2014, Thiel Fellow Vitalik Buterin invented Ethereum, which he has described as "a next-generation cryptocurrency and decentralized application platform." Ethereum is powered by a blockchain, just as Bitcoin is, but the key difference is that Ethereum was specifically crafted as an application platform, whereas Bitcoin was created primarily to facilitate bitcoin transfers. To this end Ethereum has a robust, built-in programming language that is far easier to use than Bitcoin's. Ethereum co-founder Joseph Lubin founded ConsenSys in late 2014 to be the premier venture production studio for Ethereum applications and the company has been leading many of the innovations in the space, such as the uPort identity platform mentioned above.

Traditional computer-security models rely on perimeter defenses, where all security resources are devoted to making sure that no one can permeate the outer levels of the infrastructure. But once they do—which they seem to do quite regularly— they have access to the entire system. This is what exacerbated the damage Sony experienced in their hack. Blockchain technology inherently requires that each action be permissioned with strong cryptography, so that compromising one account or one computer would only affect those particular resources.

"I have no doubt we are in the earliest days, and that blockchain capabilities will be integrated in our day-to-day lives everywhere over time. I look for real leadership also to come from the growth 'start-up' nations of the world—societies often un- or under-banked and rapidly entering the mobile revolution," says Christopher Schroeder, venture investor and author of Start-up Rising.

One example of an entrepreneur who has tapped into these opportunities is Elizabeth Rossiello, founder of BitPesa, who has been living and working in Kenya since 2009. Today, BitPesa is used in Nigeria, Uganda, Kenya, and Tanzania, working across five mobile-money networks, with instant access to over 60 banks. Elizabeth started her career at Credit Suisse and Goldman Sachs, before moving to Kenya to conduct micro-finance institute ratings and analysis across the region. While working across Africa, she became acutely aware of all of the payment challenges in the region, despite the emergence of mobile money solutions such as M-Pesa. Elizabeth saw an opportunity to use bitcoin to create a truly regional digital cash system. The problem she saw with existing mobile-money systems is that they're effectively closed-loop digital-currency systems, where each country's cellular network is exactly that: country-specific. Prior to BitPesa, there was no regional payment solution to connect all mobile-money providers. Users of M-Pesa cannot easily transact on different mobile carriers, and digital cash is owned by the individual telecom providers in each unique country.

BitPesa started by creating a local market for bitcoin in each country enabling people to exchange local currency. Then it began enabling organizational payments where employers could exchange bitcoin with BitPesa for local currency and BitPesa could deliver the local currency proceeds of the exchange to their employees' mobile-money accounts.

Last year, in an apparently anti-competitive move, Safaricom cut off Rossiello's BitPesa from its M-Pesa network. Following this move, Rossiello provided the following statement: "BitPesa does not compete directly with M-Pesa. Rather, we enable global digital transactions that build bridges between African companies and those around the world. While our first integration and corridor did connect the global bitcoin network to M-Pesa, the BitPesa team quickly added payout and collection corridors to and from Kenyan banks, Nigerian banks, and five other mobile money corridors in Uganda, Tanzania, Nigeria, and even Kenya. We have commercial partnerships with some of the biggest companies across the continent, such as Interswitch Ltd and Bhaarti Airtel. As a company founded in Kenya, we built on M-Pesa; as a pan-African enterprise, we have gone above and beyond M-Pesa."

Sam Cassatt, Chief Strategy Officer of ConsenSys says blockchain technology has many advantages over mobile-payment networks such as M-Pesa. While M-Pesa is siloed, public blockchain networks are inherently open, thus allowing additional entities to create value as opposed to proprietary systems such as Safaricom.

"If we take the open-platform, market-network concept to its logical conclusion, we could have an open backbone for a financially inclusive global economy," Cassatt says. "If people in the developing world—and the developed world—were using an open financial platform that wasn't 'owned' by a particular bank or group, additional value and financial services could emerge." BitPesa is also helping address another big problem for the region: low currency supply. Today, Nigeria, and Kenya have a shortage of US dollars. Because oil prices are low, these countries are not getting as much USD deposits in. Instead of having to use USD, these countries now have the opportunity to use bitcoin as payment and partner with the United States bitcoin exchange to cash out in US dollars.

This technology enables an entire economic region to have another way to access the global financial system, one that provides unique inherent benefits such as being open, low-cost, and instantaneous. Instead of traders having to carry large sums of money taped to their bodies because they had no way to send money to suppliers overseas, they can now use bitcoin to safely and cheaply complete these transactions. This technology could be fundamentally transformative for everyone from the consumer up through the largest institutions.

Bringing Blockchain and Bitcoin Technology to the Middle East

In the Middle East North Africa region (MENA), the percentage of adults who are unbanked is close to 80. MasterCard estimates that, as of December 2015, 85 to 90% of all retail transactions are still being made in physical currency despite the presence of payment cards and e-payment solutions. For online, digital sales, cash on delivery accounts for anywhere from 70 to 80% of transactions in the region.[6] For customers who do have credit cards, many still prefer the perceived security of cash on delivery.

Just as Elizabeth identified an opportunity in Africa, Ola Doudin saw one in the Middle East. In early 2015 she founded BitOasis, a Dubai-based start-up that created the first digital-currency payment platform and digital consumer wallet in MENA.

274 *Defining the Cutting Edge*

The wallet has more than 22,000 users across MENA and Asia, and nearly half are women. BitOasis currently has active users in the UAE, Saudi Arabia, Egypt, Morocco, and Algeria. In the future, Doudin wants to focus on other emerging markets, including Turkey and Pakistan.

Through her past experience working in the private sector, in the non-profit world, and alongside businessman Fadi Ghandour, founder of Aramex and several entrepreneurship projects, Doudin gained what she describes as a 360-degree view of the challenges in the region for e-commerce, payments, and fintech.

'Seamless' Cross-Border Payments

Ola says BitOasis is an infrastructure platform that provides tools for consumers and businesses to transact using bitcoin. The exchange and the wallet enable users to transact by using digital currency, store it in their wallets and pay merchants, freelancers, and businesses internationally. Her product offers cheaper, more accessible financial services that do not rely on traditional banking infrastructure.

The newly formed Global Blockchain Council is leading initiatives around potential pilot projects, creating proofs of conccpt, and drawing up case studies on the value of the technology. It could thus help influence and develop future regulation and policy. Doudin says the creation of the Council highlights the value of technology in the region. "Two or three years ago, a person would sit in a meeting with a bank or regional investors who would have no idea what bitcoin or blockchain is," she says.

The Council consists of 32 members, including Dubai government entities (Smart Dubai Office, Dubai Smart Government, and the Dubai Multi Commodities Centre, or DMCC), multinationals (including IBM, SAP, Microsoft), the Ethereum Foundation, ConsenSys and other blockchain start-ups (including Kraken and Doudin's BitOasis). The entity is backed by the Dubai government's Museum of the Future, a project that aims to foster the development of new technologies.

One of the Council's first pilot programs involves Doudin's BitOasis and the Dubai Multi Commodities Centre (DMCC). The pilot program seeks to solve a problem for companies licensing paperwork, and potentially streamlining payments. BitOasis is working on other blockchain proofs of concept in payments and in different applications to highlight the potential of the technology and to shape regulations.

"Now, because of the Council and the momentum and interest it's generating, I sit in meetings and people say they've heard about blockchain, bitcoin and the Council, and are eager to know more and be part of this emerging space. Even if they don't yet have a full understanding of the technology, the launch and activities of the Council did shift the view within financial services and tech on bitcoin and blockchain."

In the Middle East, banks operate from 9 a.m. to 5 p.m. from Sunday through Thursday, while international markets work on a Monday-to-Friday schedule. This means that entities in the Middle East doing business with international companies are disconnected from trading and transacting with international markets for a good part of the week. This causes delays and a loss of business, specifically if the business relies on commodities and stock trading. "If you're a business relying on trading in international markets on a daily basis, it's a huge problem because there's no mechanism for seamless and efficient cross-border payments throughout the week to keep your business running smoothly," says Doudin. "It's also a problem that's unique to the Middle East and other countries that adopt Friday and Saturday as their weekend. Since banks are off on Thursday afternoon and on Friday, even payments sent on Sunday don't get processed until Monday. We need to be able to receive international payments seamlessly, and businesses need a way to be plugged into the market when the banks are off."

BitOasis is developing a product that taps into the payment infrastructure. "Dubai is the world's hub for emerging-markets trading and movement of goods, people and capital," Doudin explains, "and we want to be right at the center of that, enabling seamless cross-border payments across those markets."

Revolutionizing Identity through Blockchain

The radical transparency of blockchain's distributed ledger means that in the future, it could be possible to create a database that reflects who's who and who owns what in a manner that cannot be altered or changed. It makes it possible to provide specific, granular access to identity attributes that are relevant on a case-by-case basis. For example, a doctor would be able to see a patient's medical records without being given access to their financial history. One company piloting this is BitNation, a start-up founded by Susanne Tarkowski Tempelhof. It offers blockchain IDs and bitcoin debit cards to refugees.

Another area of progress being pioneered by ConsenSys's uPort platform is that of identity attestations. The idea is to enable people, companies, and governments to attest to various aspects of an individual's identity, such as their social reputation, credit rating, or citizenship. The individual can decide what attestations to make public, and which to only reveal per request.

10.3 To Be or To Build: Women and Artificial Intelligence

By Tracy Saville

Pioneering artificial intelligence for the human race to help unlock our human potential, Tracy has designed, engineered, and delivered human-scale solutions in technology, transportation, housing, climate, waste management, education, and human services. She is a life-long entrepreneur, a champion of the most vulnerable among us, and a mentor to women and youth. Her past ventures include media and publishing, transformational leadership science, construction and lumber, and renewable energy.

Statistics and metrics are good. Metrics dictate what we measure, value, and design. Statistics tell us what is happening over time. Are we winning? Are we getting what we want? Metrics and statistics define our current understanding of anything—we know what we know, we know what we don't know, and data plus our personal insights and perspectives fuel our actions. This is good. Data plus smarter technology, then, can help us be even smarter. But are we getting smarter?

I look at the world we live in today, a world generally engineered mostly by men who in turn have created the current ecosystem of global investment and technology, and I conclude: the technology innovations and their value, the digital systems our lives run on, and the capital wealth our data perpetuates, are controlled by those who own it. We aren't getting smarter, because we're still designing and investing in technology that lives in an old world.

Most of those who own the world of money and technology have inherent biases, because all people do. Even the best data can limit our ability to be smarter if bias keeps us from changing the metrics or building new technology paradigms so we can ask new questions. There's a reason why Silicon Valley is slow to change, why experts suggest it will be 118 years before we achieve gender equality in economic terms (pay parity) as a human race.[7]

People have no reason to change their behavior if the wealth they control is working just fine without it. If men make up the majority of those people, it doesn't take a statistical genius to see that we need a different tack. This is what the *Internet of Women*^{sm} is doing: engineering and funding the new tack—the new investment-and-technology ecosystem, free from the binds of unchecked bias.

It's compelling to note that the data of the world is sufficient to show us that in the case of capital and technology, we have a massive diversity and gender problem. And much like climate impact, we also have a big enough

10.3 To Be or To Build: Women and Artificial Intelligence

body of data to show us the associated costs, problems, and challenges we're facing, and thus why we have to change the status quo. Yet, we didn't see big oil lead the global sustainability change. Nor will we benefit by waiting for Silicon Valley regulars to change their status quo.

Y Combinator (YC), which has statistically been responsible for most of the unicorn valuations in the last eight years, has funded more than 1,000 start-ups, with a total valuation of $65 billion, including Dropbox and Airbnb. Since 2008, only 11% of total applicants funded had female founders in their mix. YC co-founder Jessica Livingston[8] spoke with Bloomberg's Emily Chang in April 2016 about unicorns and the role YC had in making them. In this interview Jessica was also announcing her plan to take a year's sabbatical to be with her kids. She is a primary founder of YC. I was intrigued by the disparity of parity in the YC stats because of her influence in the design and development of YC. I wondered about her taking a year off to get to know her kids better, although I can commiserate, as most women with children would: what will happen now that she was stepping out, leaving a virtual gender and diversity vacuum in her wake.

If we look at the facts about women's impact in corporate and investment returns, the delta between how few women share in that $65 billion of market share as compared to what the data AND conventional metrics tell us about the impacts women have on bottom lines, can't logically be attributed to oversight. Bias seems to be getting in the way of even the establishment's best interests. Yet, even this hasn't so far been sufficient to move the needle farther or faster.

Consider how women are outperforming their male counterpoints in ROI:

> "Women-led private technology companies are more capital-efficient, achieve 35% higher return on investment, and, when venture-backed, bring in 12 percent higher revenue than male-owned tech companies." (Women 2.0) 2013[9]
>
> "Companies with women in top leadership positions are more financially successful than companies run by men. ... Companies with strong female leadership generated an average return on equity of 10.1 percent per year, compared to an average of 7.4 percent for those without top women leaders. The study, which examined more than 4,200 companies between December 2009 and August 2015, is in sync with a September 2014 report from Credit Suisse that also found a link between companies with more female executives and higher returns on equity—as well as higher

278 *Defining the Cutting Edge*

> *valuations, stronger stock performance and higher payouts of dividends. The Credit Suisse report was based on a database that tracks the gender mix of about 28,000 executives at 3,000 companies in 40 countries.*"[10,11]

The irony of a world today that shows women outperforming their male counterparts in ROI as compared to how little they get funded is that women are the primary catalysts for changing the metrics and re-designing ROI values. So it is up to us to find a way to fund and use technology to do that. The *Internet of Women*[sm]? Necessity is always the mother of invention. Welcome to the age of equality engineering, or what I fondly like to refer to as the human, intelligent web, the web that is being re-engineered by women. This is where I believe the next wave of value will come from in technology in the world.

Let me introduce you to a few more women who are building that world.

Google "women and artificial intelligence." First notice this search result: "Inside the surprisingly sexist world of artificial intelligence..."[12] Okay, well that's not surprising actually, given how rarely women are referred to as doing disruptive work in this space. This is not to say that women are not disrupting in AI. I am. Others are.

This *is* to say it is not sexy to cover the disruptive work of women in this space. And unless women are at the mouths of media channels that do the reporting about disruptions or at the heads of capital pipelines that invest in disruption, it's like a tree that falls in the forest with nobody around to hear it. Money and the gatekeepers of the definition of what's valuable or relevant set the headlines.

The third article in your Googling experiment will land you on a list of women in engineering who are changing computing. I list them at the end of this chapter because they can't be found in such a grouping anywhere else on the web.[13] These women and their contributions are changing the world. They are working in the science of cognition, machine learning, and leading-edge forms of artificial intelligence—computer scientists, digital sociologists, visionary entrepreneurs who get it—but have you heard of them?

What about the Opportunity Statistics?

Women's entrepreneurship is, and will continue to be, a leading force for socio-economic development for the new world we're growing into, a world defined by our ability to collaborate and work together.[14] According to a 2015 report by McKinsey Global Institute, advancements in women's equality could result in an increase of global GDP by $12 trillion by 2025. In a world where the "full potential" of women were unlocked, this number rises to as much as $28 trillion.[15] We have to design for and fund that potential.

That opportunity will be unlocked by The *Internet of Women*^{sm}.

It will also be unlocked by women such as Women Moving Millions founder Jacki Zehner, a former Wall Street wunderkind. She's using her wealth and experience to rally other women of means to fund new enterprises beyond Silicon Valley. She is among my favorite leading examples of the new age of women and money.[16]

Still, while some of the world's most influential women and the Internet of Women are using their wealth and minds to change the world, pointing their billions and innovation at peace and shared value rather than sitting at tables of economic systems that exclude equality as a value system, we have to suffer the ignorance of bias on our way to building and setting our own tables.

What We're up Against

In March 2015, a friend sent me an e-mail from a former venture capitalist who suggested why going with women investors would be a bad idea. He said money, successful business models, and so on come from a select few places, and if we fail, I won't get bonus points for my approach in working with irregular money partners. He made a point of saying that I needed a recognizable male in my domain so that Wall Street and Silicon Valley, who control the game, would take us seriously. He said we needed to do it the way Silicon Valley would recognize. Otherwise, even if we got funding, our valuation would never be as high as if we did pass the gauntlet, such as the Y Combinator gauntlet.

He said that, having worked with Andreessen Horowitz and other tier-one firms is the only way to go to get to the valuations you want. He also said much of what a great male venture capitalist does is help develop a CEO, thus implying that there were no female CEO mentors who could adequately mentor me, a woman, and that I must need mentoring because what woman doesn't? Where will my next job come from? Certainly not my wealthy girl pals, because only the world's leading venture firms can do that.

Finally, he suggested that who I selected to lead my funding was very important and that tin-cupping wealthy women creates many points of light AND confusion, because women of means are not VCs, and they would have little oversight of me as the CEO. In other words, unless it's a male venture capitalist running the show, a recognized male-accepted lead on funding mentoring me to success, my company, My Swirl, won't be legitimate, and maybe it will even be irresponsible, because unless a male venture capitalist has me under his foot, I can't be trusted to get us where we need to go.

It's hard enough to solve the complex business challenges of building and funding disruptive artificial intelligence that has the force to change the world. When I calculate the running tally of how much gender and cultural bias costs me—from a founder's perspective—my trajectory goes from hard to expensive. Time lost to market to navigate bias costs in ROI, opportunity costs in the market lost because I have to create the funding ecosystem to fund my own venture before I get there, or the gyrations my team goes through to stand up, and still have to work 10 times harder to qualify because we're women? It's very expensive.

Every moment that people don't have the technology solutions of all the women in this book and beyond, is one more moment we have to wait to get past bias so we can get to a world without it. The real cost is borne by humanity. This is why I am enamored with machine learning, with artificial intelligence (AI), and the idea that AI can help us to sort out our own blind spots and to accelerate next-level wisdom. AI in the hands of the *Internet of Women*^{sm} is a powerful trend. If we can evolve emotional intelligence and awareness faster, we can engineer equality into the world. If we can do that, we can engineer and design for values of collaboration, better human relationships, and a "we" world. We can change the metrics. We can change the data. And we don't need anyone's permission to do it.

An interesting irony to consider: Google's Julia and Apple's Siri—notice how men have given AI female attributes. It seems that women can BE AI, because supposedly we're less intimidating than men, but we can't qualify to BUILD AI, because that's the domain of men. Or if we do build it, we can't get it financed, or have it be taken as seriously as we would if we had had the blessing of Silicon Valley, or the "right venture-capital" firm, or a man on our team who had thought it up in the first place.

Leigh Alexander, in her piece *"The tech industry wants to use women's voices—they just won't listen to them,"* has an illuminating take on this.

> *"It's fitting that our modern fiction about AI should go hand in hand with horror-laced tales about men's failure to correctly estimate women. Increasingly, AI helpers from Apple's Siri and Microsoft's Cortana to talking home thermostats, GPS and fitness apps default*

to a female voice, as lots of research suggests that both male and female consumers prefer it. *The likely explanations of this are many and probably driven by social conditioning—we want our virtual assistants to seem pliant and non-threatening, competent but not domineering.*

Maybe AI development is also influenced by the geek culture ideal of being alternately serviced and encouraged by a hard-earned digital princess—the nostalgic science fiction fantasies of white guys drive lots of things in Silicon Valley, so why not the concept of AI? ... No matter the reason, the voices of women—and their creators' ingrained concepts of modern womanhood—are leading the AI frontier. But we have no reason to believe that the male-dominated technology industry understands us at all."[17]

10.4 Opportunities for Women in the Green Economy and Digital Infrastructure

By Tess Mateo

Tess Mateo is a Special Advisor to the United Nations (UN) Global Compact on Gender and Climate and the Founder and Managing Director of C^X Catalysts, which develops public-private partnerships in clean energy, water, sustainable food, infrastructure and health. She helps female entrepreneurs access wealth-creating opportunities in the high-growth green economy and digital infrastructure sectors.

In 2015, leaders of over 190 countries signed both the UN Agenda 2030 for Sustainable Development and the Paris Climate Agreement, promising a global transition to greener and more-inclusive growth. Recognizing that today's inequitable and carbon-emitting methods of economic growth are unsustainable, leaders in the public and private sectors are seeking transformative ideas. My vision is for women to own and/or control at least 30% of the world's resources within the next century. It may be an audacious goal given that women currently own or control less than 1%. But, short of 30%, women will not have enough influence in major decisions impacting the environment, the economy, or civil society to drive sustainable growth.

There is a growing understanding that women are under-utilized assets who can positively contribute to sustainable development. Research studies by

McKinsey Global Institute, Ernst & Young, Deloitte, and the World Economic Forum have quantified women's unrealized value at between $12 trillion and $20 trillion. Government leaders made gender equality and women's empowerment the fifth Sustainable Development Goal (SDG) of the UN Agenda 2030. However, a greater paradigm shift in the near term is needed beyond giving women equal rights and opportunities in order to truly transform our current unsustainable growth.

Women need to attain the majority of wealth generated in the coming decades in order to adequately gain control of 30% of the world's resources. To accomplish this, they need to become immersed in extremely fast-growing sectors. Fortunately, these opportunities do exist. Sir Richard Branson, Founder and CEO of the Virgin Group, has said, "Climate change is one of the greatest wealth-generating opportunities of our generation." The World Business Council for Sustainable Development (WBCSD) estimates in their Vision 2050 report that market opportunities for products and services that help reduce harmful greenhouse gas emissions will be between $5 trillion and $10 trillion per year in 2050. McKinsey and the Ellen MacArthur Foundation find that over $1 trillion a year could be generated by 2025 if companies built up circular supply chains to increase the rate of recycling, reuse and remanufacture.

The digitization of economies and societies is another exponentially expanding area with wealth-creating opportunities that female entrepreneurs should pursue. Professor Klaus Schwab, Founder and Executive Chairman of the World Economic Forum, believes we are at the beginning of a Fourth Industrial Revolution. The first three revolutions mechanized production through steam power, enabled mass production with electricity, and created digital capabilities that have connected billions of people. The Fourth Industrial Revolution advances new technologies to combine the physical, digital and biological worlds, this blurring the real with the technological world to create cyber-physical systems. New information-technology (IT) ecosystems are emerging that disrupt sectors such as energy, agriculture, banking, healthcare, and transport. For example, in 2017, Google will introduce its fully autonomous (self-driving) vehicle. It uses an integrated system that includes IT networks, sensors, global-positioning systems, data analytics, and machine-to-machine learning. According to Lux Research, self-driving cars are expected to create $87 billion worth of opportunities for automakers and technology companies, especially software developers. To take advantage of this growth, we must raise the number of women pursuing opportunities in information, communication and technology (ICT).

Looking at these fast-growing areas suggests that perhaps the greatest wealth-generating opportunities in the near future may be where they connect, and even overlap, thus creating multi-faceted technology solutions that help address climate change. For example, when governments around the world pass new clean-energy mandates resulting from the UN Paris Climate Agreement (the largest, most comprehensive global climate deal), companies and governments will need existing and new technologies to meet the governments' planned contributions. Technology plays an increasingly important role in allowing companies to properly comply with governments' reporting requirements, as seen in global financial reporting after the Sarbanes Oxley (2002) and Frank Dodd (2010) regulations. Compliance with any reporting requirements needs analytical tools and automated processes. Therefore, Fourth Industrial Revolution technologies, involving complex algorithms that transform previously unstructured data into usable and traceable forms, will decrease human error by streamlining the reporting process through automation. Technological solutions help lower the operational and reputational risks of organizations that have to report any information. The challenge is how to get more women and girls interested in and equipped for starting businesses in these fast-growing areas, even if they are not in positions today to even learn about these trends. Women and girls need more than inspiration and adequate training and support to be successful green entrepreneurs. Based on my own personal experience, they also need male champions and sponsors to provide access as well as guidance. As governments and companies go digital and transition to low-carbon growth, women can earn a greater share of the wealth created, and can go farther than their predecessors, who were limited by a glass ceiling.

To help expose women to these opportunities, my firm, C^XCatalysts, committed, along with the International Federation of Business Professional Women at the UN's historic Conference on Sustainable Development (UNCSD, Rio+20), to empower ten thousand women in green-economy businesses. For example, we designed public-private partnerships to create women-owned micro-franchises with multinational companies (MNCs) such as General Electric and Greif, and multilateral institutions such as the International Labor Organization (ILO) to deliver clean water and energy to underserved populations. For our GE-Habihut solar water kiosk project in Nairobi, Kenya, the ILO estimated that, beyond the 36 female entrepreneurs who provided 5,000 people with purified water with GE's filtration technology, another 150 women vendors had benefitted from the well-lit informal marketplace that the solar-powered Habihut kiosk had helped create. An interesting outcome was the transformation of these women from unknown community members to sought-after and well-respected productive citizens once they controlled a desired resource (clean water). Ten percent of the global population lacks access to safe water, according to Water.org, and this small project shows the tremendous potential for women when they access and control critical resources.

With the Internet becoming part of the essential infrastructure for the 21st century, C^XCatalysts is currently exploring how best to develop women-owned green-technology companies. As a first step, we are piloting women-owned Internet service providers. With very few women-owned or women-controlled digital infrastructure, this can be an exciting and impactful development, especially as the demand for closing the digital gap and cloud services grows and new Internet-based services are introduced.

What Led Me to This Ambitious Journey?

If anyone during the World Wide Web's birth in the 1990s had suggested that I would someday champion women in ICT, my friends and colleagues would have laughed. Women's empowerment was not a priority when I was climbing the corporate ladder.

During the Internet's exciting early days, everyone—male and female—wanted to join that decade's new economy, and no woman I knew complained of gender discrimination. In fact, I left a highly coveted management-consulting position to become number two at a business-to-consumer start-up when a college friend recruited me for one of his investments. The company was able to survive the dot.com meltdown by nimbly shifting the strategy to focus on our online performance-incentive platform. I learned valuable lessons about flexibility and real-time data measurement and analytics during my more than 20 months riding that exhilarating rollercoaster.

When a mentor became the global CEO of PricewaterhouseCoopers (PwC), one of the largest professional-services firms in the world, he recruited me to rejoin PwC. During his two terms as global CEO, we experienced nine of the top ten corporate accounting scandals of all time, including Enron in 2001, and Lehman Brothers and the global financial crisis in 2008. Working at a high level in yet another industry experiencing tectonic shifts, my dot.com rollercoaster was nothing compared to the accounting profession's earthquake: namely, the passing of Sarbanes-Oxley legislation, which expanded US public-company reporting requirements. With tremors felt around the world because of the far-flung operations of companies listed in the United States, those who understood global accounting and auditing standards and processes saw their value rise. It was during this tumultuous time, while working at the very top of a 165,000-person, $31 billion global organization, that I became more cognizant of gender discrimination.

Representing my boss on his CEO non-client commitments, I often found myself to be the only woman at the table. While I had been aware of the dearth of high-level women, I had not embraced this as an issue until I had experienced it firsthand. While supporting the restructuring of the American and global financial markets and rebuilding trust in them, I learned how to navigate the complexities of driving private- and public-sector collective action. After joining the World Bank's Private Sector Leaders Forum, formed to engage the private sector in advancing women's economic empowerment, I saw how the passion of just a few individuals could drive lasting change around the world by harnessing the power of global institutions. It was then I realized I could do something.

The WBCSD's Vision 2050 helped me define what that something could be: a transformative vision based on a new perspective on women's economic empowerment that aligned with my capitalistic background. Vision 2050 was an 18-month project working with over 200 MNCs that represented 14 industries and 20 countries—an undertaking that required a deep immersion in climate science, energy scenarios, geopolitical risks, and research on sustainable consumption and production. Since the fastest economic growth was expected to come primarily from developing countries, corporations and governments learned that their interests were converging. Companies, impatient to wait for underfunded governments to educate their citizens or build infrastructure, invested in developing these new markets. Further, there was a collective recognition that gender equality, according to the UN and the World Bank, significantly contributes to advancing economies and sustainable development. The 200 MNCs agreed to 40 "must-haves" to achieve their shared vision of 9 billion people living well within the limits of the planet by 2050. The five priorities that most interested me were women's economic empowerment, public-private partnerships, renewable energy, clean water, and infrastructure solutions. When PwC estimated that Vision 2050 would generate $8 trillion of business opportunities based on the collective group's inputs, I thought women should win some of it and knew that I could help.

My new journey was further defined when I was recruited to join the International Advisory Board of the Global Summit of Women and to co-chair the Women in the Clean Built Environment for the International Trade Centre. Since I was frequently asked to make introductions and translate between government ministers and corporate executives, I realized my value was in connecting people and ideas. While becoming a mother and thinking about my children's future may have initiated my transition from ambitious restructuring consultant to caring sustainable-development advocate, the demand for my connecting skills provided a means to drive women's economic empowerment into the fast-growing green and digital economies. My firm's name, C^XCatalysts, speaks for itself as we aim to spark opportunities by connecting concepts and communities to leverage change.

What Can People Do?

If global leaders who support the United Nations (UN) Sustainable Development Goals or World Business Council for Sustainable Development's (WBCSD) Vision 2050 believe that working towards gender parity is necessary for sustainable growth, then women owning and controlling at least 30% of the world's resources should be a goal. To accelerate women's progress, institutional and cultural norms must be adjusted and aligned to this goal. Girls and women need to be inspired and trained to pursue entrepreneurial opportunities that are not only high-growth with high margins, but also valuable at a strategic level to their country's social, environmental, and economic-development interests.

How to Attract More Female Entrepreneurs into the Green-Economy and Digital-Infrastructure Sectors

Governments, companies, academic institutions and non-profits must inspire and educate girls and women to pursue entrepreneurial opportunities in high-growth, high-reward sectors such as science, technology, engineering and math (STEM). More visible female role models are needed in these sectors so women and girls can see the possibilities. Broadening female roles in films, books, educational materials and toys so that they include women who are developing new technologies or addressing climate change will inspire girls at a young age, helping build a pipeline of future female green-economy tech entrepreneurs. For example, the Geena Davis Institute on Gender in Media is helping influence the portrayal of female characters on TV and film, whether through different story lines or otherwise. The non-profit SheHeroes also develops videos about inspiring women who are thriving in STEM careers for educators.

Educational institutions at all levels should complement their female-directed STEM initiatives with entrepreneurship training (or vice versa) so

10.4 Opportunities for Women in the Green Economy and Digital Infrastructure

that women and girls can develop the skills to thrive in the green-tech sectors. Business organizations, government agencies, academia and women's and girls' networks helping implement the Paris Climate Agreement, the UN Agenda 2030, or both should be challenged on how they can individually and collectively best contribute to growing the ranks of women green-economy tech entrepreneurs. For example, the United Nations Conference on Trade and Development (UNCTAD) operates its entrepreneurship program, Empretec, in 36 developing countries. UNCTAD plans to add to its existing women's business award more entrepreneurship training for women, including modules on climate-change actions and mitigation strategies.

To determine which green economy sectors to pursue, women's organizations can partner with organizations such as Sir Richard Branson's Carbon War Room (CWR). CWR endeavors to accelerate the adoption of business solutions that reduce carbon emissions in the order of gigatons and advance the low-carbon economy. CWR looks for innovative business models and brilliant new technologies, and works to harness the power of entrepreneurs to implement market-driven solutions to climate change. CWR has specifically identified the following green economy opportunities: renewable energy, recycling and waste reduction, energy efficiency in buildings, and building materials.

To highlight opportunities in waste, C^XCatalysts helped to launch the Reuse Opportunity Collaboratory (ROC Detroit) in conjunction with General Motors, Fairmount Santrol, Detroit's Economic Growth Corporation, and the United States Business Council for Sustainable Development (US BCSD). ROC Detroit is a platform for Detroit industries, government agencies, small and medium-sized businesses, academics and entrepreneurs to create beneficial environmental, societal and economic opportunities from Detroit's underused material and waste streams. For example, Carla Walker-Miller, Founder and CEO of Walker Miller Energy Services, which designs and implements energy-efficiency solutions for utility, residential and commercial clients, now uses GM's excess auto insulation for her projects. Once ROC Detroit is adequately funded, US BCSD's award-winning Online Material Marketplace can be deployed to multiply these types of matches.

Another way to increase the number of women in green-economy technologies is for companies with relevant products to develop female entrepreneurs as distributors. For example, Schneider Electric, a French multinational energy company, provides vocational training to more than 30,000 people in South America. They created a women's program in Brazil to train energy professionals for both rural and urban communities. The women sell and maintain Schneider's new renewable-energy home solutions.

How to Help Women-Owned Green Economy and Technology Firms

Governments implementing the UN Agenda 2030 should, at national, state and local level, institute purchasing (procurement) policies to source from women-owned firms. Public-sector spending accounts for 30 to 40% of GDP in developing countries, and 10 to 15% in developed countries. The International Trade Centre Global Platform for Action on Sourcing from Women Vendors aims to increase the amount corporations and governments spend on purchases from women vendors. While the United States had set a target in 1994 to award 5% of their contracts to women business owners, it took over 20 years to achieve that target. The governments of Samoa, the Republic of Uganda and Republic of Rwanda have joined the initiative. According to the World Bank and the Clinton and Gates Foundations, supporting women business owners has positive ripple effects, including alleviating poverty because women reinvest more of their earnings than men in their children and communities.

The private sector can do the same. The benefits of supplier diversity and inclusion are well established: General Motors and McKinsey Global Institute say that supporting women can be a way of broadening a company's reach and market intelligence. According to WEConnect International, women own one-third of all private businesses, but earn only 1% of corporate and government procurement spending. To help address this, the Clinton Global Initiative worked with Vital Voices and WEConnect International to launch "Advancing Women-Owned Businesses in New Markets," where 24 leading companies, including Boeing, ExxonMobil, IBM, Coca-Cola, Pfizer, Walmart and Marriott committed to increasing their spending on women-owned businesses. Between 2013 and the end of 2014, spending on women-owned businesses exceeded $3 billion, surpassing their three-year target in just one year. The United Nations Women Empowerment Principles (WEPs) encourage companies to expand their business relationship with women-owned enterprises. UN Women and the UN Global Compact have already recruited over a thousand of the world's leading companies to sign the WEPs.

In developing countries, entrepreneurship is, more often by necessity than by opportunity, dominated by women entrepreneurs. These small women-owned businesses are initially considered part of an informal or grey economy, which includes activities that are not regulated or protected by the state. However, when these businesses become successful, their ownership often moves to the men. Access to finance and difficulties in formalizing their

10.4 Opportunities for Women in the Green Economy and Digital Infrastructure

businesses because of laws and regulations are some of the challenges that women face and that need to be addressed.

In addition, while access to markets, financing and legal ownership is critical, supplier readiness is just as important, so that women business owners can learn how to navigate the procurement process. Companies, government agencies and women's organizations must collaborate across the small and medium enterprise development lifecycle to identify, develop, and scale high-potential female entrepreneurs who can become strong suppliers.

Who Is Doing It?

There are success stories out there that give me hope that women can achieve the 30% goal. At the Meeting of the Minds 2015, I met Emily Kirsch, Co-Founder and CEO of Powerhouse, the world's first and only incubator and accelerator dedicated to solar energy. Located in Oakland, CA, the incubator houses the most ingenious solar start-ups and the accelerator invests in them. With solar equipment and hardware at historic lows, Powerhouse has become a leader in incubating technology companies that help solar providers innovate and deliver their services. It was founded in 2013. Its first two start-ups were Mosaic, which helps finance residential solar installations in the United States, and Powerhive, which partners with utilities and independent power producers to provide microgrid electricity solutions to rural communities around the world. The United States solar market is expected to grow 120% in 2016, according to the U.S. Energy Information Administration. With the August 2015 passing of the Clean Power Plan, which aims to reduce carbon pollution from power plants that spew 40% of carbon emissions in the United States, renewable energy use, including solar, is projected to soar. In addition, according to Bloomberg New Energy Finance's market research, global solar demand will grow from 2% in 2012 to 18% in 2030. Serving this high-growth sector, Emily, Powerhouse and the companies they incubate are well-positioned for significant wealth creation.

Further, Myriam Maestroni is the Founder and CEO of Economie d'Energie SAS and is equally impressive. We were on a panel together at the 2014 Global Summit of Women held in Paris. Myriam's company, based in France, develops innovative programs to promote energy efficiency in all sectors. After leading several European energy companies such as Dyneff, Agrip and Primagaz, she identified an opportunity to help companies and governments meet the European Union's ambitious plan to reduce 20% of its primary energy consumption by 2020. Recognizing the need for sustainable energy suppliers to change their focus from supply chain, operations, and safety to customers and their energy use, she launched Economie d'Energie. The firm helps businesses with programs and technology to teach their customers how to use less energy. Their "web-life-web" programs, which are designed and hosted online to help consumers renovate their homes in compliance with France's efficient-building policies, have benefited over 350,000 French households by the end of 2015. Their 45 digital platforms have already attracted five million visitors, and they predict that another two million will join in 2016 alone. Clients include France's largest companies, such as Auchan, Leclerc, Carrefour, Schneider, Rexel, Total, GDF, and Esso. With buildings generating a large portion of the world's carbon emissions, Myriam's company is also well positioned for growth.

There are other success stories out there. We must find them, encourage that success, and use it as seeds to grow further successes.

While I would love to see women own and/or control 30% of the world's resources in my lifetime, based on historic trends, it usually take 100 years for such large cultural shifts. Maybe there are ambitious girls out there who can develop technologies to accelerate this ☺!

10.5 CyberSheroes: The Inspiration Behind CSI

Professor Dr. Mary Aiken is a cyberpsychologist, and Director of the CyberPsychology Research Network. CyberPsychology is an emerging field within applied psychology, focusing on Internet psychology, virtual environments, artificial intelligence, gaming, digital convergence, mobile telephones and networking devices. Mary's research focuses on virtual profiling and cyber behavioral analysis, specifically youth behavioral escalation online. The CBS prime-time show "CSI: Cyber" is inspired by Mary's work as a cyberpsychologist.

Dr. Aiken answers questions sent to her by the editorial team.

1) Please explain the emergence of cyberpsychology, and how you got involved and ultimately became a leader in the field.

The scientific study of cyberspace began in the early 1990s, when researchers first attempted to analyze and predict human behavior mediated by technology. Cyberpsychology itself is about 15 years old as a discipline. It studies the impact of technology on human behavior. The primary focus is on Internet psychology but we also consider virtual environments, artificial intelligence (AI) and intelligence amplification (IA), gaming, digital convergence, and mobile/networked devices. Zheng Yan, the editor of the *2012 Encyclopedia of Cyber Behavior*, has predicted that the subject will enjoy exponential growth due to the rapid acceleration of Internet technologies and the "unprecedentedly pervasive and profound influence of the Internet on human beings." There are now over 30 journals publishing around a thousand papers a year in this subject area. Professor John Suler is widely acknowledged as the founder of cyberpsychology.[18]

I think that cyberpsychology can make an important contribution to developmental psychology—I have just published a paper on the subject. It considers the impact of digital or interactive screen time on the developing infant (aged 0 to 2 years). Very few people know that the American Academy of Pediatrics (AAP) recommends that media use in this age group should be actively discouraged.[19]

The area of cybersecurity can also benefit from cyberpsychological insights—my specialist area is forensic cyberpsychology, which focuses on criminal aspects of behavior manifested in a cyber context. There is currently a lot of interest in what is described as "Human Factors" in cybersecurity—and cyberpsychology can certainly help to illuminate this space—the intersection between humans and technology.

2) It's very interesting when you talked about being measured on performance versus gender, and how performance may be a great equalizer. Could you elaborate?

I am often asked what it feels like to be a woman working in areas traditionally dominated by men. To be honest, when it comes to my work I don't really have a construct of gender—I have a construct of performance. My favorite quotation is, "Always dream and shoot higher than you know you can do. Do not bother just to be better than your contemporaries or predecessors. Try to be better than yourself." That's from William Faulkner.

I know that there are many challenges for women—inequality is a major issue. In particular, science, technology, engineering and math areas have always attracted and retained more boys and men than girls and women. However, what has worked for me in my career to date is not focusing on, or worrying about, my gender, but focusing on and enhancing my performance.

3) How did you get involved in the show CSI?

I was working on a White House led research initiative that was focused on exploring technology solutions to technology facilitated human trafficking—we presented the results in the West Wing of the White House in 2013. There was some media coverage, which brought my work to the attention of the entertainment industry in Hollywood, this in turn led to an invitation to meet CBS network executives in Los Angeles. I was scheduled for a 15-minute interview with the then President of CBS Entertainment, Nina Tassler, this brief session actually turned into a two-hour discussion, and shortly afterwards I was invited to become a producer on a new show *CSI: Cyber*. The show is inspired by my work as a cyberpsychologist. My character Avery Ryan, is a Cyberpsychologist FBI Cyber Crime Special Agent, and is played by the Oscar-winning actor Patricia Arquette, who is tasked with solving crimes that *"start in the mind, live online, and play out into the real world."* I am very proud of the fact that my character Avery Ryan is the first female lead in a CSI show.

First and foremost I am an academic, an educator who cares very deeply about the impact of emerging technologies on humankind. My aim is to deliver insight at the intersection between humans and technology. The Hollywood part just happened, I cannot claim that I planned it. I was simply the right subject matter expert, in the right place, at the right time. That said, it was an incredible experience to become part of a talented creative writing and production team, to generate scripts and cybercrime stories that would reach out to a worldwide audience, and would build on the legacy of the CSI franchise.

The show provided a platform to raise problem issues associated with technological developments—I do worry, however, that we are more or less "sleepwalking our way into an age of technology"—I often pose the question as to why we rush to adopt each new emerging technology with the collective wisdom of lemmings leaping off a cliff. As I point out in my new book *The Cyber Effect*—what is new is not always good—technology does not always mean progress. I think we need to stop, reflect, and ask ourselves the following questions; Have we had the conversation about what we want in cyberspace? Have we had a greater societal debate? What is the role of governance? What is the role of good practice or cyber ethics? We know what we want from real world society—what do we want from cybersociety?

Every time I give a talk or interview I try to point out that technology was designed to be rewarding, engaging and seductive for the general population—did anybody really think about the impact on criminal, deviant or vulnerable populations such as children? As women I think we are well placed to tackle technology facilitated problem behavior—we are very aware of our protective instincts—especially when it comes to the vulnerable—I firmly believe that we have an important role to play regarding all things cyber.

4) What advice would you offer to those who want to get into this area?

When it was aired in March 2015, the pilot of our show *CSI: Cyber* broke a Guinness World Record when it was broadcast in 171 countries simultaneously—for an academic that is an incredible global platform to educate and entertain regarding cyber safety and security. It has also facilitated the introduction of my discipline cyberpsychology to a worldwide audience—I think far more people watched the show than would ever read a paper in an academic journal.

5) Do you have any specific advice or recommendations for women looking to enter this field?

I would highly recommend that young women pursue a career as a cyberpsychologist. It's a very flexible and timely qualification, and enables the pursuit of a career in academia (research), or clinical practice (e-therapy), or in industry (digital marketing). My advice is to focus on the science and on performance, be the best that you can be, and most importantly enjoy yourself, and remember: If you are passionate about what you do, it will never feel like work.

10.6 Women in Cyber: Filling the Gap

"Solving new and creative challenges requires as much creative and diverse thinking as possible. Those who think they can solve these challenges with less than half the planet's participation are just bad at math."

By Dr. Alison Vincent

Dr. Alison Vincent, Chief Technology Officer, works with Cisco customers large and small to understand how technical solutions can help drive business. After completing her PhD in cryptography Dr. Vincent's career has spanned three decades incorporating research, strategy execution, product management and business development. She works with Cisco to help inspire young children in the potential of technology.

Cybercrime is set to become the UK's most common offense overtaking any other kind of crime in the United Kingdom (UK). The UK's National Crime Agency released its National Strategic Assessment, which noted that losses from such crimes exceeded £16 billion annually, already making up a significant proportion of what the UK loses from organised crime.[20]

Despite growing demand and the tremendous opportunities in the job market, cybersecurity remains an area where there is a significant shortage of skilled professionals regionally, nationally and internationally, regardless of gender. *"The International Information Systems Security Certification Consortium estimates that last year about 332,000 InfoSec pros joined the global workforce of about 3.2 million, but the field needs as many as 2 million more practitioners"*[21]

But it is frustrating that, even in this field of security, the representation of women is alarmingly low.

I found myself asking: Has this always been the case? What about all those women who worked at Bletchley Park? So before we look at the current situation further and share some personal journeys, let's roll back the clock and see whether history has something to teach us.

Cyber in the Past

The Bletchley Park codebreaking operation during World War 2 was made up of nearly 10,000 people, of whom about 75% were women. That in itself is a striking statistic. At the time, the majority of women in the United Kingdom did not have a bank account. Much of the work was repetitive, and involved just crunching out the numbers. But there were a few remarkable women who are finally being recognized as cryptanalysts who worked at the same level as their male peers:

- Mavis Batey (formerly Lever)
- Margaret Rock
- Joan Murray (formerly Clarke)
- Ruth Briggs

Mavis had great language skills and a capacity for logical thinking. Margaret was a graduate mathematician from Bedford College London (which became part of Royal Holloway College, where I studied cryptography). Both Mavis and Margaret joined Dilly Knox and his all-female team in the "Cottage" and broke many a cypher. Joan, another mathematician from Cambridge (who is represented in the film "*The Imitation Game*") eventually became the Deputy Head of the "Hut." It was quite rare for a woman to hold a leadership position. And then there was Ruth, a language scholar from Cambridge, who we are just beginning to learn more about.

But these stories of women at Bletchley have remained, much like the work itself, shrouded in secrecy. The main focus has been on the male professors who dominated the top levels. To find information about these ground-breaking, inspirational women, you really have to dig deep.

My other blast from the past came on a recent trip to Bletchley for an off-site meeting that was being held in the Museum of Computing. The Museum houses a fantastic collection showing the advances made in computing through the ages, complete with the machines that helped to break the Enigma Code. But there was one beautiful piece of mechanical engineering—the oldest working computer. As it performed a computation, the valves lit up in a hypnotic way before the machine spit out the answer. That wasn't the most fascinating find, though. It was an old black-and-white photograph that was displayed in front of the working computer. It was the last group of computer scientists who were completing their research—and it was 50% women.

Personal Journey

Let's start to make things a bit more personal. It's about time I outlined my own personal journey in this domain. Are you sitting comfortably? Along with many young people, my decisions around my academic choices through my school years were very much driven by what I found interesting or easy! In the UK, by the time you turn 18, your decisions around your A-level choices tend to filter you into one or another university degree on your way to a career. This fact in itself is not highly recognized nor communicated to the school leavers that I come across even today. My A-levels were an eclectic mix of mathematics, physics and geography.

After my A-levels it was time to choose my degree. This was before the days of gap years or the choice of not going into higher education (and the high level of financial debt that young people nowadays find themselves saddled with on entering the workforce). As I mentioned, I chose mathematics (as I found it easy)—but thank goodness I decided to include computer science, too. PCs were not widely available at the time (yes, I am that old), and I had noticed that some of the "lads" at school had started to play with Spectrums.

I didn't really have any understanding of what these things were—but I did realize that it was the start of something new and incredibly important, and I wanted to be part of that.

As my three years at university unfolded, as I had suspected, the computer-science curriculum became more developed. But the turning point for me was a professor, Fred Piper, and a module that was offered to the mathematics undergrads in our third year. This module was intriguingly entitled, "Codes and Ciphers." Because I had spent many a long hour, when I was younger, de-ciphering messages in children's quiz books, signing up for this module was a no-brainer. Professor Piper enlightened us about the wonders behind the Enigma machine at Bletchley Park, the simplicity of the Caesar Cipher, the practicalities of the one-time pad, and finally the concepts of public- and private-key cryptography. I was hooked!

As the finals came round, discussions started in parallel about what job to take after the BSc. Along with many others, I entered the UK undergraduate hiring phenomenon called the milk round (don't ask me how it got its name). This was a program where all major UK employees advertised their opportunities in their companies, and the best and the brightest wandered around their stalls trying to make sense of the job titles that were being described. After investigating many a company, I was in the final stages of accepting a job with a major confectionery company on their generic fast-track management-development program.

It was at that moment when my results were announced. Without wishing to boast, I attained a First Class Honors Degree, and there was much celebration at the Mathematics Department farewell drinks. During the fun in the sun, my Professor took me to one side and asked me a very simple question: "Alison, have you ever considered doing a PhD?" After picking up my jaw from the floor, I honestly answered that I hadn't. I am of the opinion that not many women think they can—and they need that friendly nudge from someone who sees more in them.

That conversation changed the course of my career. I started my three-year commitment to a PhD in Mathematics and Cryptography, under the guidance of Professor Piper and my industrial sponsor, Professor Chris Mitchell from Hewlett Packard. I found being close to industry really valuable. It ensured that the research remained grounded, and it also gave me visibility in what it was like to work full time in a software-development organization.

It's amazing how quickly time goes by. Before I knew it, the time had come to make more choices about my future. With the connections that Fred had, it wasn't long before a UK Government Agency (which shall remain nameless—but I'm sure you can work out which one I mean) approached me for an interview. I duly went along and learnt a very valuable lesson in life. It's really important to know what you don't like. During the interview it was all male and too many of them were wearing white socks and sandals, and I made the decision there and then that I couldn't work in that kind of environment. Instead I chose a career in software engineering, working for a large corporation.

I have since spent a very happy 25 years working for multiple software companies, in a variety of roles, from software-engineering leadership to business development and mergers and acquisitions. If these roles appear to be on the periphery of cryptography let me show you how they are intimately related. I was always involved in:

- Reviewing patent applications in the field
- Ensuring security issues were part of design reviews
- Making sure the development organisations were educated on coding for security

Even in my current role as CTO for the sales function, I have had the sales-support engineers take part in the Cisco Security "Ninja" program.

If I were to pass on the lessons I have learned from my journey, I would summarize them as follows:

- Keep doing what you love and find easy!
- Make the most of chance conversations—you never know where they can lead.
- Find the people who really know what you are capable of—and who will remind you of those skills.

On an individual level, we all need to continue staying informed about the new developments, products, and standards for this field, because these things change on a daily basis. New technologies become available every day, and part of our job is to be aware of them and see how they apply to where we are now.

Cyber in the Media

No one should ever underestimate the power and influence of the media. I began to ask myself, what kinds of role model are out there for women in cyber? Are they truly aspirational, or are they in fact closing off this avenue to a potential career option?

Cybersecurity is often viewed as a hacking discipline, with visions of a "geek" sitting alone at a computer well into the night, breaking into something. You only have to tune into the likes of WarGames, Swordfish, Antitrust, "Live Free or Die Hard" to see the nerd in action. Hackers and The Net both tried to introduce female role models, but both Sandra Bullock and a young Angelina Jolie tend to be on the screen more as eye-candy than because of their intellectual capabilities—though Angelina was the inspiration for Keren Elazari, who now speaks on Hackers as a force for change in the Internet world.

Only more recently have we seen worthier interpretations. Let's take three examples.

1. The Imitation Game: Based on the story of Alan Turing and his efforts to hack the Enigma code. Here the cryptographer, Joan Clarke is finally portrayed on screen, by Keira Knightley. I personally felt they attempted to downplay any hint of glamour or particular femininity. Joan was there for her intelligence and higher ability than the other male crossword completers at the interview.
2. The Da Vinci Code: We are introduced to Sophie Neveu, a police cryptographer. The undertones of the movie itself are littered with the balance between male and female, with Sophie proving to be the yin to the yang of Langdon. The male and female work together toward a goal without the female's being subordinate to the male in any way. Sophie is quick witted, agile, caring, compassionate and brilliant. She also has a PhD from Royal Holloway College—like me!
3. CSI: Cyber. Here, special agent Avery Ryan works to solve crimes as a CyberPsychologist for the FBI. Not only do we get a view of the extent of cybercrime outside hacking, including cybertheft and the introduction to the dark net. We are also introduced to a new kind of role—a psychologist who also has cyber skills—and to a "hack for good" culture. The Avery Ryan character is based on the real-life pioneering psychologist Professor Mary Aiken—another fantastic role-model.[22]

So maybe the media are finally turning things around? Certainly when I see large players such as the BBC in the UK invest in online programs such as "Make it Digital", I think we are at the start of something great.

Examples of Prominent Careers in Cyber

There are many options for a career in the cybersecurity arena, and not all of them are about being able to de-cipher complicated mathematical codes or sitting geek-like at a terminal hacking for good or bad. The interesting thing about this area is that adding gender has nothing to do with gender at all per se. Attackers, hackers and crackers in the past were mostly male; the folks who outran them, hunted them, and shut them down were also male. The "battle" or "cyberwar" was fought, waged, prosecuted, and so on.

Now, we have diverse crooks and a wide expanse of humanity inadvertently breaking things and clicking things and sharing things they perhaps should not be sharing. The bad "guys" are starting to look a lot like the population. The problem solvers are not. The scale is different, the way things work is different, and what is deemed good enough to get past the broken bits and to balance "safe" against "creepy," or "minimally viable" against "perfect", is different.

In short, solving new and creative challenges requires as much creative and diverse thinking as possible. Those who think they can solve these challenges with less than half the planet's participation are just bad at math. Its takes everyone, in a variety of roles, to produce great cybersecurity, and the career options are vast. Consider the process behind how a web application is created and launched. It requires the efforts of software developers to do secure coding for handling data on the Internet, network engineers and security analysts to handle the flow of data on the Internet, and security researchers to come up with appropriate technologies and cryptographic algorithms and products to ensure the security of the data while it is in rest and in motion. And then there is the deployment of the application itself, for which various devices and services such as firewalls, Intrusion Prevention systems etc. Here is a quick, non-exhaustive list of options:

- Geopolitical analyst, to bring non-data-driven thinking to the cybersecurity picture, such as human motivations, nation-state entanglements, and diplomacy
- Risk specialist
- Information-security (InfoSec) investigations and program management
- Security sales, pre-sales technical and consultancy roles
- Process and audit roles
- Policy and education roles
- Business/data/IT analyst, architecture, product owner, quality-assurance analyst, scrum master, content manager
- Penetration testing, vulnerability discovery, exploit writing, incident response, malware detection, intrusion detection and network hardening

Career options for women are exactly the same as they are for men—unless it involves being able to bench-press 200 lb! And I don't see that in the list above.

Interestingly, there is one role where women are pulling their weight and that is in Governance, Risk and Compliance (GRC) with "*20% of women identifying GRC as their primary functional responsibility, compared with just 12.5% of men holding similar positions. GRC is one of the fasting growing information security roles where women tend to dominate, the report said, with women typically possessing key character traits that enable them to succeed in GRC roles.*"[23]

But it's still critical to see more girls and women participating, and understanding the importance of the cyber world and how it's foundational to all jobs in the future. Do we really expect artificial intelligence to be developed by just the one gender or ethnicity *and* to meet the needs of the entire global population? As we develop intelligence within devices, diversity becomes crucial to our social development, so getting girls involved at a young age is just one way of ensuring that that happens. Technology is already a part of our social fabric.

Let me share with you a story from a female IT security operations manager. It starts when we are small. I have a son and a daughter. My son started playing soccer (which is traditionally a guy's game in the UK) at age 4. My daughter wanted to do what her brother was doing. So as soon as she was four, we got her some boots and sent her to soccer practice. There were no other girls playing, but the coaches had said it was open to all, and they loved that a girl wanted to join. A friend of my daughter's asked her mom if she could also play. The mother said 'No, dear. Look, they are all little boys—it's a boys' game.' And there's your problem.

So a call to action. If you are in the field already, you have a responsibility. You need to go into schools and talk to classes about the huge variety of careers on offer in this area, and the importance of the topic. This way we can begin to break the stereotype that cybersecurity is all about geeky guys writing code and hacking into systems.

References

[1] https://blog.ethereum.org/2016/07/27/inflation-transaction-fees-cryptocurrency-monetary-policy/
[2] https://letstalkpayments.com/50-of-women-globally-are-financially-excluded-and-fintech-can-change-it/
[3] http://www.wsj.com/articles/the-revolutionary-power-of-digital-currency-1422035061
[4] http://www.coindesk.com/santander-blockchain-tech-can-save-banks-20-billion-a-year/

[5] http://www.ibtimes.com/facebook-one-out-every-five-people-earth-have-active-account-1801240
[6] http://www.gsma.com/mobilefordevelopment/wp-content/uploads/2016/04/SOTIR_2015.pdf
[7] http://reports.weforum.org/global-gender-gap-report-2015/press-releases/
[8] http://www.bloomberg.com/news/videos/2016-04-08/y-combinator-s-focus-on-female-founders
[9] http://www.bloomberg.com/news/articles/2013-02-20/women-who-run-tech-start-ups-are-catching-up
[10] https://www.shrm.org/hrdisciplines/diversity/articles/pages/women-ceos.aspx.https://www.msci.com/documents/10199/04b6f646-d638-4878-9c61-4eb91748a82b
[11] https://publications.credit-suisse.com/tasks/render/file/index.cfm?fileid=8128F3C0-99BC-22E6-838E2A5B1E4366DF
[12] qz.com
[13] Geekfeminism.wikia.com
[14] Minniti and Naudé 2010
[15] http://www.mckinsey.com/global-themes/employment-and-growth/how-advancing-womens-equality-can-add-12-trillion-to-global-growth
[16] http://www.womenmovingmillions.org/events-and-programs/programs/speaker-programmembers/jacki-zehner/
[17] http://www.theguardian.com/technology/2016/mar/28/tay-bot-microsoft-ai-women-siri-her-ex-machina
[18] See my forward on John Suler http://www.cambridge.org/ae/academic/subjects/psychology/applied-psychology/psychology-digital-age-humans-become-electric?format=PB
[19] Haughton, C., Aiken, M. P., & Cheevers, C. (2015). Cyber babies.The impact of emerging technology on the developing child. Psychology Research. 5(9), 504–518. DOI:10.17265/2159-5542/2015.09.002
[20] http://www.scmagazineuk.com/cyber-crime-overtakes-physicalcrime-in-the-uk/article/445014/
[21] http://searchsecurity.techtarget.com/opinion/Women-in-cybersecurity-The-time-is-now
[22] See also the interview with Professor Mary Aiken in Section 12.1
[23] https://www.isc2cares.org/uploadedFiles/wwwisc2caresorg/Content/GISWS/2015-Women-In-Security-Study.pdf

Manifesto

By Laurie Cantileno & Rahilla Zafar

The Internet of Women (*IoW*) Manifesto aims to clearly spell out key recommendations from the book as guidelines of how to alleviate gender disparity in the tech world. Let's be clear: by no means do we want to exclude men – the *IoW* in and of itself is an inclusive collective mind. There's been an incredible amount of research on how diversity brings more options, creativity, and solutions. Everyone needs to bring more than just the talk they profess and do their part like the many individuals we highlighted. This Manifesto includes key recommendations drawn from this book. You must change the way you attract and retain women or you risk continuing with business as usual and ultimately deepening the rift. Here are 11 succinct ways in which we can collectively move the needle:

1. 21st Century Benefits: The Workplace Has Changed. Why Haven't You?

Technology is rapidly improving the potential of the world to communicate faster and more efficiently. Unless there are solid business reasons (other than just keeping an eye on your employees from a stance of trust) for daily face-to-face communication, allow some flex time and work from home. You will find people actually work harder and longer when given the freedom.

Design professional events from an inclusive perspective—not just sports events or happy hours. Use a mix of event types and venues to draw in many kinds of employees with sensitivity to employees who have families. Take pulse surveys to find out what interests employees. As we think globally, be conscious of religious or cultural restrictions, i.e. work events centered around alcohol, late-night hours leading to problems with transportation, or restrictions around men and women being alone together.

2. Invest in Lifelong Growth: Mindfully Invest in Your Female Employees

Employees are companies' biggest asset. Companies can't grow solely with a bunch of ideas from only executives and no one to execute and scale them. Since women bring many talents to the table, you'd be missing out if you didn't create a workplace that invests in them. Help unlock their potential and it's your gain. Attracting women is one thing, retaining them is another challenge. You may not think you have what it takes to be a mentor, but you'd be surprised. These are the people who inspired you at some point of your career, who reached out, who recognized your potential. They gave their time and advice to help you get to where you needed to be. Mentoring is a special and powerful relationship. It is important to realize that it is reciprocal. Mentors get a lot out of mentoring; personal satisfaction as well as a realization of themselves.

In our book, we highlight Elizabeth Isele's work from Babson College who works with corporations and governments on retiring the world "retirement" and advocating for an "Experiential Economy". Multigenerational teams must be the new norm for the 21st century as ageism is a factor. A key focus point of her finding is reverse mentoring where digital natives are paired with experienced professionals where they both learn new skills from each other.

Another example of comes from Naila Chowdury, chair and cofounder of Alliance4Empowerment. As the former CEO of Grameen Solutions Limited, a tech company founded by Dr. Muhammad Yunus, recipient of the 2006 Nobel Peace Prize, Chowdhury is now piloting a program in San Diego that promotes opportunities for girls and women through 'social businesses.' Working with a demographic of women, many of whom are victims of human trafficking, she says that many lack the confidence they need to lead and be in charge. Additionally, many women are shielded from being involved in finance, which she incorporates into their education, she explains: "Through providing a social credit, we built a bridge between micro-credit and social business. Micro-credit is less than $1,000. We are achieving social credit empowerment through technology by providing Internet access, tablets, and mobile phones along with the business so we have direct access with the lenders and the investors have direct access with the borrowers."

3. Men and Women Need to Highlight the Role Models of Women

Women need to actively and consciously create opportunities for other women. More female investors should mean more funding for female entrepreneurs. More female entrepreneurs should mean more women on teams.

More women in high-growth businesses provide those role models and should make it easier for other women to enter the workforce. It's women supporting other women that will push us beyond the tokenization of women in organizations.

"I know one way I can help is by ensuring women hold leadership roles in my start-ups. We know that a lot of startup founders come from start-ups themselves. They go build experience working at a start-up, and then leave and do it themselves with the right networks in place." – Davis Smith, Founder of Cotopaxi and baby.com.br.

"With more data at hand, it became clear to me that the problem was cultural: women were trying to compete in industries created for, and run by, men. On one hand, they had to face all sorts of biases and extra hurdles and on the other hand, they were not as adapt as male colleagues in navigating the system (either because of access or because of upbringing/socialization). Moreover, with no interesting role models at the top, leadership didn't feel attainable or desirable for many of the women I met." – Ravi Karkara, UNWomen

4. Working Together to Inspire Girls Early: Parents, Schools, Corporations

There is much to be said about getting girls started early. The formative years are where self-confidence is developed. From the earliest age it begins with encouraging parenting. Parenting that treats girls as equal to boys. Parenting that provides children with both female and male professional role models. Parenting that recognizes gender nuances and enhances those areas where children excel regardless of gender. Schools need to make sure their staff gender is equally balanced and fosters an environment where children are not intimidated to ask questions. Schools have a responsibility to promote mathematics and the sciences for both boys and girls without preconceived notions of their capacity for absorption. Supporting girls in these fields is imperative until the gender gap is closed. It is amazing to experience and watch pioneers and women who break glass ceilings, but at some point, parity will be the standard.

Anyone can have an influence by serving as an "executive sponsor" as Edna Conway, an executive at Cisco Systems where industry leaders, neighbors, and experts contribute to educational systems and support programs focused on youth from ages 5 to 13. She draws attention to the Cisco Girl Power Tech Program, which is also part of the UN's International Girls in ICT Day. This is an example of community outreach where corporations can get involved in improving community resources in areas neglected by governments; where young girls would have to otherwise stay home if the basics like water stand in the way of their education.

In addition to large corporations, government ecosystems can make a difference as well. Al Anoud Faisal Abdul Rahman, a prominent female angel investor, and Professor Muhammad Rahatullah Khan of Effat University studied four female entrepreneurs in Saudi Arabia. Three of them cited assistance from the Chamber of Commerce and banks, and utilized government initiatives that provided substantial assistance along with support from their families. Additionally, all four start-ups received business and entrepreneurial training to enhance their technical education (their CEOs were computer science graduates). Two of the women credit coaches and mentors who were seasoned entrepreneurs that had gone through similar challenges within growing their businesses.

There are many examples in the book on the significance of role models, as well as many call outs for entrepreneurs and corporations to walk the talk by promoting, financing and highlighting women in order to create these role models.

5. Network Broadly, Creatively, and through Personal Connections

The "good old boys" not only have a club but they have a network. One thing that is reiterated over and over again in many of the stories is the concept of community and the benefit of networking, which is something that many women are only getting to realize.

Sallie Krawcheck, Chair of the Ellevate Network has said, "Networking has been cited as the number one unwritten rule of success in business. It is a good part of what begins to separate the pretty successful from the very successful on the second leg of your career . . . and beyond."

6. Reward Desired Behaviors and Structure: Incentives to Foster Change

People appreciate being recognized for good behavior while showing that you actually value them. Ensure employees have time away from the office to spend on themselves and with their families. Balance is key in life and this should be rewarded as well. Encourage employees to actually take vacations and personal time off and do not bother them while they're gone.

Davis Smith of Cotopaxi speaks of an incentive structure he set up within his start-up so employees are rewarded when taking vacations and having the opportunity to volunteer within a non-profit within or outside the company's community. This is certainly something larger companies can do as well by creating a 'digital currency' that measures and rewards staff for not working during off hours for example and rewarding their employees for it.

7. Men Are Important Allies and Advocates.

We can use all the help we can get, so we are encouraging the men to join in. Men championing women is much more effective than women doing this alone. Since men hold most of the prominent positions in the world this is essential. Men need to educate other men as well.

In Nicole Merl's case when she was looking to launch #WomenVote, she wasn't afraid to ask Thomas Cook, someone who she knew had the skill set and contacts to help make her idea a success. For men, don't be oblivious, and think about how it might feel to be a minority within a room full of a dozen people who all look the same. With the help of Thomas's valuable partnership support, her proposal for #WomenVotes was approved as the first-ever virtual co-op authorized by Northeastern University to research, develop, and implement a digital social impact project as both an experiential and innovative learning experience. Their purpose is to engage women from all political affiliations to share their voices, empower #WomenVotes, and make a difference through a #ServeAmerica portal filled with non-partisan resources, candidate information, and volunteer opportunities.

Investing in human capital whether in the public or the private sector, often ends up supporting women the most. One example of this is Robert Mosbacher Jr.'s career. Robert became President and Chief Executive Officer of the Overseas Private Investment Corporation (OPIC) in October 2005. As head of OPIC until 2009, he oversaw the investment of billions of dollars in emerging markets around the world, and helped the organization evolve from just providing microfinance loans (the majority of which went to women) but also larger lines of credit targeted towards small to mid-size enterprises (SMEs).

8. Be Conscious about Unconscious Bias

Remember that there is "Unconscious Bias" where someone may be overlooked as a result of preconceived notions and not treated in the same manner because they do not seem to fit the model. Don't be oblivious be consciously active in your awareness.

Think about how the other person might feel to be a minority within a room full of a dozen people who all look the same. Understand unconscious bias where someone may be overlooked and not treated in the same manner because they do not seem to fit in. One example of this comes from Impact Enterprises, where despite there being a close and supportive community, women still struggled in a public group setting. Although Zambia is a patriarchal society, support groups, whether Lean In in the US or something similar to what Impact Enterprises launched in Zambia, are an important safe space for women, where they feel comfortable and welcomed building their communication skills and discussing ideas without being judged.

9. Women Advocating for Women

It's no longer the "nice" thing to do—it's your obligation as a woman. Yes, that is a bold statement. We are not here for the shrinking violet approach. Pull a female through. Pull many females through. It's been done like that for hundreds maybe thousands of years in the male community.

This is another inclusive edict. In our chapters you have seen many men as champions of women. In our investment chapter in particular, there's an emphasis on men and women investing and supporting women in the entrepreneurial ecosystem. The saying of the 'good old boys club' exists for a reason. It is time women stopped competing with other women and came together to support women. When one woman supports another we are collectively creating opportunities for each other as a whole.

10. Seek to Find a Sponsor and a Champion versus simply a Mentor – What's the Difference?

Quackenbush credits finding a champion who not only encouraged her but was willing to go out on a limb and advocate for her. *We are challenging readers to not only find someone to sponsor you through mentoring but to do the same for someone else.*

Sponsorship is an act not an activity, where you champion an individual. Sponsorship is very different from mentoring, however, and the acceleration exponentially increases when both are pulled through. Sponsorship is also something that can be measured. So be a sponsor or find a sponsor; that person who is willing to go out and advocate for you to get a promotion or progress within a company. And when you find that sponsor, always be positive around them. Remember they are going out on a limb advocating for you; respect that trust and instill confidence so they continue to be your champion. Then pave it forward to someone else. Recognition is one of the most productive ways to retain employees, especially women. Look for those high-potential candidates and actively pull them in. Since there are so many less women in technology than other careers, it will take significantly more effort to seek them out and recognize them. More women retained means more women role models, which means more women will follow suit.

Another example of this comes from Serpil Bayraktar, a Principal Engineer in Chief Technology and Architecture Office at Cisco Systems. Early on in her career, she felt as though she was experiencing "explosive growth," but after having her first child, she felt she had reached a plateau in her career. Even though she raised her children as a single mother, things began to change in her career trajectory when she had a manager whom she describes as being unusually supportive of women in the workforce, particularly those with families. "He immediately asked me to take the lead on a list of projects, and invited me to the table to voice my opinion in setting the direction of our team and in making major decisions," she says. She credits these opportunities with helping her develop into a Principal Engineer. She was also given support in creating a new "Women in Technology" program at Cisco. In addition to her mentors' believing in her, she says they opened up their networks to get her engaged with a supportive set of people who would become her advocates.

However, you cannot ask for sponsorship, this only comes through trust and relationship building. This is why mentorship and networking are such critical steps in this journey, and also closely tied with pulling someone through from point #9 above. These are the people who advocate for you and stick their neck out voluntarily.

11. Men and Women Must Support Female Entrepreneurs and Investors

The need for not only more women investors but also male champions have been highlighted. One such example is Hala Fadel, an accomplished female entrepreneur and now an investor. Hala is a partner at Leap Ventures, a late-stage venture-capital firm based in Beirut and Dubai. Her passion for advancing entrepreneurship opportunities has inspired her to be an angel investor in start-up companies in the Middle East and Europe. She's also launched a co-working space in Beirut called Coworking+961. Recently she founded MIT Technology Review Arabic, the Arabic edition of the MIT Technology Review Magazine, which covers science and innovation in the Arab region.

She's also founder and chair of the MIT Enterprise Forum of the pan-Arab region, an organization that promotes entrepreneurship and that organizes, among other things, the MIT Arab Start-up Competition. It is one of the biggest entrepreneurship contests in the region, and it engages entrepreneurs from 21 Middle Eastern countries in training, mentorship, exposure and networking.

An interesting set of figures emerges from the MIT Arab Start-up Competition's 2014 survey. According to it, women make up 24% of applicants in the competition and 21% of semi-finalists. Seventy-one percent of the semi-finalists have teams where at least one female has a leading role. And as much as a quarter of female applicants submitted tech projects, as compared to half of male applicants. "There is definitely a groundswell of interesting women-led or founded projects. We just need to discover and nurture them, and provide them with the assistance to grow," says Chris Schroeder.

Chris is an American entrepreneur and CEO who invests in consumer-facing innovation. He's also a well-known advocate of global entrepreneurship, writing, and mentoring start-ups, all over the world. He's the author of the book, *Start-up Rising—the Entrepreneurial Revolution Remaking the Middle East*, which is the first major analysis of start-ups in the Arab world. For the past four years, he's supported Fadel, attending the regional MIT competition to judge and mentor start-ups.

"If we look at how gender inequality is being addressed, we see that women are increasingly using platforms to engage, teach, collaborate, and build. As an illustration, the largest per capita user of YouTube in the world is Saudi Arabia, the largest demographic is women, and one of the largest content areas is education. Play this out globally over time and there is zero chance women's engagement in this phenomenon does anything but increase exponentially," he says.

Closing

Accelerating Change: The Internet of Women's Role in Catalyzing Global Gender Equality

By Bobbi Thomason Ph.D.

The Internet of Women documents a rich array of lived experience in the technology sector and the visions of women and men for achieving gender equality. The stories collected in the book provide readers with a narrative about the state of women in today's digital economy, providing case studies, personal accounts, and solutions for moving this Internet of Women from awareness to practice. Despite the dearth of women working in and leading organizations in the technology industry, these recollections and hopes offer cause for optimism and inspiration.

Female scientists, technologists, engineers, and mathematicians worldwide are breaking barriers and making incredible contributions to their fields. While we have not yet reached gender parity in terms of pay or leadership representation, the writings and case studies collected in this book document that there are exciting cultural shifts taking place around the globe. From Deemah AlYahya, the first female Saudi executive at Microsoft, to the impact of the UN's emphasis on girls and technology education in the SDGs (Sustainable Development Goals), to the increased female labor force in Zambia, a policy change that was inspired by the MDGs (UN Millennial Development Goals), *The Internet of Women* captures stunning examples of progress from around the world.

The authors tackled the immense question of what a global, inclusive movement to gender equality looks like, within and across our national borders, sharing numerous examples of how change is being ignited around the world. Michelle Lee, the first female Under Secretary of Commerce for Intellectual Property and Director of the United States Patent and Trademark Office (USPTO), reflects on her Girl Scout badge in sewing and cooking and how it inspired her to create an IP badge that exposes young women to innovation and the process of invention at an early age. Malala Fund co-founder Shiza Shahid shares her background of mentoring young women in Pakistan and how she's now working to direct more investment to women innovators around the globe.

To conclude this volume, I share some insights from existing research on gender equality and evidence-based solutions on how women can overcome gendered barriers in technology.

Why Create an Internet of Women: Economic Effects of Gender Inequality

As the authors of this volume have argued, research shows that gender inequality is not only a pressing moral and social issue but also a critical economic opportunity. The gender gap in representation and leadership in the technology industry, and job market more generally, occurs at a cost not only to women but also to the teams, families, organizations, and countries in which women work and live. There are numerous studies that show the value of diversity in boosting productivity and the bottom line within all levels of a company, from entry level to the boardroom,[1] as well as the critical role women play in enhancing the collective intelligence of groups.[2]

There is also a growing body of research on the business benefits of specifically having women in the top leadership roles of organizations. Fortune 500 companies, for example, with at least three female directors experience at least a 53 percent increase in their return on investment capital compared to those with fewer or no women directors.[3] In a study of 1,500 American firms in the S&P, female representation in top management also improved financial performance for organizations where innovation is a key piece of business strategy.[4]

In September 2015, the McKinsey Global Institute (MGI) released their report, "The Power of Parity: How Advancing Women's Equality Can Add $12 Trillion to Global Growth," in which they explore the economic potential available if the global gender gap were to be closed. Their research found that, in a full-potential scenario in which women play an identical role in labor markets to men's, as much as $28 trillion, or 26 percent, could be added to global annual GDP in 2025. This impact is roughly equivalent to the size of the combined American and Chinese economies today.

Just as women are underrepresented in organizational leadership and positions of authority, they are over represented elsewhere, such as unpaid work. In comparing husbands and wives in the United States, who are both employed full-time, women typically provide 40 percent more childcare than the men.[5] This overrepresentation of women in childcare responsibilities exists across the globe. In Australia and Italy, women do twice as much childcare as men. In Japan-five times, and in India-ten times. The male-female disparity in domestic housework responsibilities is equally large. In the United States, a woman does 30 percent more housework than her male spouse. In Australia and the Netherlands this percentage jumps to twice as much and is three times greater in Japan.[6]

The chapters of this book and leading social science research are also aligned in the conclusions that there is great variation in the specific hurdles and contexts in which women navigate their careers and lives. Parental leave is one such topic. The United States is the only developed country that does not mandate paid maternity leave. The issue of paid maternity leave leads to unique challenges and perceptions. In the United States, many families fall into poverty as the result of unpaid leave after the birth of a child. In the Middle East, where many countries mandate maternity leave benefits, women are often perceived as "more expensive," making companies more reticent to hire them.[7] Take another example. In the Netherlands, 76.6 percent of women work part time. While the Netherlands has been lauded for offering women work-life balance at a national level, there is a paucity of Dutch women in senior leadership roles. Interestingly, removing part-time workers from the analysis, women working full time achieve management roles nearly as often as men working full time.

Not only do these brief examples speak to the range of experiences of women globally, but they also highlight that no country is uniform in its achievement or failure of gender equality. Within each country, the experiences of women will vary by race, socio-economic class, and more.

How to Conceptualize the Internet of Women: Taking a Global and Intersectional Approach

As conversations about gender equality continue around the world, it will be critical that we continue to take a global and intersectional approach. Debates about gender equality in the United States illustrate how, even well-intentioned conversations may not cohesively address all individuals. With this debate front of mind, I commend this volume for conceptualizing an Internet of Women that is global and that includes all women. I would not suggest that the United States is the "default" for organizations, work, or women's experiences. Yet, if we take a global approach to gender equality, there are important insights from the experience in the United States to keep in mind regarding how we build a movement of change.

In examining the feminist movement in the United States, many laud the positive impact of efforts to level the playing field, while others note a lack of inclusion inherent in them. Some have accused the current women and work conversation as "trickle-down" feminism—an expression of concern that additional wealth of already privileged women in the United States and abroad will barely benefit women of lower socio-economic status. Others have pointed out that the conversation seems to generalize the experiences of white women to *all* women.

Data reminds us that the experience of women depends on women's multiple identities. Results of an experimental study by researchers at Kellogg School of Management examining the simultaneous impact of race and gender on leadership outcomes revealed, for example, that dominant Black female leaders did not create the same backlash that dominant White female leaders did. This suggests that tensions such as the "likeability-competence" trade-off may resonate and be navigated differently by White women and Black women in America. And while the study shows that Black women may not suffer the same penalties for dominant behavior, this does not mean that they do not suffer other types of penalty—and perhaps double penalty—for making mistakes on the job, for example.

Race is only one of many dimensions along which women's experiences vary. As the examples in this book highlight, a woman's gender is a never experienced in isolation. She experiences being a woman simultaneously with her race, socioeconomic status, nationality and profession, among other identities. The concept of "intersectionality," coined by Kimberly Crenshaw, allows us to address these multiple identities together and reminds us that there is no "one-size-fits-all" feminism.[8]

To take an intersectional approach to women globally, one has to consider the national context, which can also shape how signals about gender diversity are interpreted. For example, a study that compared the United States and France found that women in the United States tended to talk about gender diversity in terms of numerical ratios of men to women, while women in France talked about gender diversity as understanding how to manage diverse groups. Accordingly, French women's interpretation of gender diversity signals were less influenced by attitudes about affirmative action.[9] As American-based diversity strategies are exported to different regions of the world, companies need to be thoughtful about how the local context will influence the ways in which such programs are interpreted and experienced. Attitudes towards and meanings of diversity depend on an individual's belief system and larger power dynamics such that the same diversity initiatives could resonate with some people and exclude others.

National context, however, is only one of many dimensions along the range of women's experiences that has only grown. The examples in this book have also included class, industry, marital status, and parental status. These are important identities to consider that women experience along with their gender. Research has even documented how in globalizing economies being categorized as "global" or "cosmopolitan" versus a "local" will also impact individuals' opportunities for professional development and career rewards.[10] While there is much work to be done in order to understand how all of these identities play out in the world of work, this book offers both accounts of lived experience and fodder for further inquiry.

What the Internet of Women Includes: Change at the Organizational and Societal Level

The chapters of this book are filled with not only stories of individual experience, but evidence of and calls for organizational and societal changes. Both of the individual and the broader structures in which individuals live and must continue to be examined. Research shows that gender inequality is a function of both biases at individual levels and patterns at social and structural levels. Consequently, we are going to need solutions at both. Popular conversations may debate if women should "lean in"[11] or if the system is so broken that women just "cannot have it all."[12] I fully agree with both sides of the argument and advocate that we need change at both the individual and institutional levels. The women and leaders featured in this book offer suggestions for first steps that we can take on both important levels.

On an individual level, research shows that biases impede our path to gender equality and that the mental models that get in our way need to change. Psychological research has documented how powerfully and instantly people categorize individuals as male and female. Several studies conclude that sex provides the strongest basis of classifying people, trumping race, age, and occupation in the speed and ubiquity of categorizing others.[13] While male traits are associated with leadership and professional authority, for example, female traits are not. These associations often occur without people even being aware of them.

This research begs many of the same questions that authors of this book have worked to solve: What can organizations do to solve this inequality in both mental associations and leadership representation?

The individual and organizational barriers are connected. In order to create a fair hiring and performance evaluation process, companies need to implement practices and procedures that block bias. For example, research has also found that when men and women negotiate for higher salaries, women more so than men are found to be less likely to be hired.[14] To prevent such dynamics, companies will need to develop clear criteria for evaluation and compensation to apply the same standard to everyone. In ambiguous situations or when criteria are unclear, people tend to rely more on stereotypes in their evaluations.[15] Research has shown that when people are held accountable for their decisions, such as explaining why they chose one candidate over the other, they scrutinize their decisions more, which also reduces bias.[16]

At the moment, there is essentially no evidence that suggests we can train people to be unbiased, so organizations hold a responsibility for putting processes in place to rein in biases. Technologies are being created that can address this. Textio, founded by Kieran Synder, for example, seeks to remove gender bias from job descriptions and can be applied to other text. Their software analyzes text for how well their words and phrases resonate with men and women. It can tell you that the word "rock star" will draw more male job seekers and can propose that instead you use more gender-neutral language, such as "high performer" or "guru." Even if we cannot stop bias, practices like this can block bias from negatively impacting our organizations.[17]

These dynamics are important across the board in organizations and especially important to women in technology. Research on the attrition of women in STEM has found that workplace culture is a significant factor in why women leave. A 2008 study that examined the reasons male and female engineers left the field found that men were more likely than women to leave to pursue advancement opportunities. Women engineers were more likely than men to report leaving because of a negative work climate.[18] A recent analysis of women in engineering that compared women who left engineering with women who stayed found that women who left were more likely to point to organizational barriers. Such barriers included hostile climate, inadequate training and development opportunities, and a lack of advancement opportunities. Notably, in the United States, women of color reported less supportive work environments.[19]

It should be noted that both men and women will benefit when we shape how our organizations conceive and reward quality work. Numerous studies, for example, have pointed to the difficulty women have in meeting both work and family responsibilities.[20] But men are also placing an increasing value on their personal lives. A Pew survey found that almost half of fathers (46%) are concerned that they are not spending enough time with their children.[21] Flexible work comes in many forms such as part-time work, reduced work hours, working from home, and taking time out of the workforce altogether for a period of weeks, months, or years. Men and women can both benefit when companies rethink how work is done and how careers progress. Thankfully, parental leave is one area where real change is beginning to happen. In 2007, Google increased paid maternity leave from 12 to 18 weeks (and paid paternity leave from seven to 12 weeks) and achieved a 50% reduction in the rate at which new mothers leave the company.[22]

However, policies alone will not create fundamental cultural changes. Individuals must enact these changes.

Organization leaders, for example, need to model new ways of working, including flexible work schedules, as it signifies that it is a welcomed and accepted thing to do. Supervisors have immense influence on the effectiveness and use of work-life policies. I believe, for example, that getting more senior men using flexible work policies would go a long way in creating change. All too often, employees do not make use of flexible work practices because of unsupportive supervisors and because they fear they will be penalized for doing so. Unfortunately, research shows they are right. Both men and women can be penalized for using flexible practices. Documented penalties include declines in pay, a lack of promotion opportunities, and mistreatment.[23] Organizations need to implement different policies, and leaders must use these policies and show that individuals who use them can thrive in their organizations.

How Can the Internet of Women Catalyze Change?

The task of achieving global gender equality is extraordinarily complex. Tackling challenging dynamics in our organizations and families—internalized by stereotypes and pervasive in national policies—will require multiple and ongoing changes. I hope companies will take responsibility and make serious changes. However, there is also great hope and opportunity beyond our organizations. The Internet offers opportunities to ensure that the varying perspectives of diverse women are heard and documented, while also connecting women by what unites them.

As the pages of this book suggest, the Internet has the potential to drive change—and to drive it quickly—by providing opportunities to store and share data, collaborate, and connect. Access to information—and the means to collect it—may be the greatest opportunity of the Internet of Women. Much (though, not all) of what drives gender inequalities today is not explicit discrimination but implicit beliefs that many people are unaware that they hold, such as associating men and not women with leadership. Research indicates that stereotypes are less virulent when women can detect and resist them. The Internet offers women a means to inform themselves about stereotypes and their potential effects. The aggregation of important data will also help the world see and understand the richness and diversity of women's experiences.

The Internet offers opportunities to coordinate and collaborate for work. Platforms, from Airbnb to Task Rabbit, for example, are at the core of the sharing (or "gig") economy, in which technology facilitates and monetizes the sharing of particular tasks ("gigs"), services, and goods. This part of the economy is currently estimated at $26 billion and growing every day. It is composed of well-known companies, such as Uber and Etsy, and smaller start-ups, such as HireAthena. While some of these platforms have been at the center of controversy, I would argue online platforms should not yet be ruled out. These platforms are malleable—women can use their voice to make them work for them, and platforms will have to listen to women's preferences and aspirations. If Uber has adjusted to local markets—creating cash-based transactions in Nigeria and offering tuk-tuk rides in India—why not customize additional features should women ask for them? New companies are also in the working to create opportunities for virtual teams and virtual organizations, which may offer novel access to employment.

The Internet also provides social means through which to coordinate and connect. Social networks, from Twitter and Facebook to Pinterest and Instagram, are low-cost, low-barrier-to-entry forums for women to tell their stories and catalyze change. An example of this can be seen through the rise in the Iranian Women's Movement, created in secret and fostered through various online platforms.[24] These platforms have helped garner support from women across social classes, enabled the movement to be as self-sufficient as possible (remaining independent from funding streams and physical locations), and provided a progressive outlet for women that challenges the current political structure.

Of course, the flexibility to either have anonymity or identification is key to creating an Internet of women and for women. In Saudi Arabia, as many as 70 percent of women use Facebook under aliases. In the United States, Facebook has been a platform for women to talk about their recovery from sexual assault. Social media and Internet networks have also shown to provide connections that have staying power and grow over time, e.g., professional organizations where women are able to continually receive ongoing advice and mentorship from interactions that may have started out face-to-face but continue online. There are also many sites that connect female entrepreneurs with mentorship, expert advice and solidarity between communities of women doing similar work.

The solution to gender equality is not simple, and the Internet will not be a silver bullet or panacea. The Internet is perhaps a space where unequal gender structures are repeated and recreated, or it could be a space in which new paradigms bring empowerment and inspiration through women's experiences. The contributors to this book believe that governments, civil society, and corporations must work together to continue to push for the cultural shift that is needed to spur innovation and support economies globally. They also believe that the Internet will have a central role in achieving the above. The next chapters of individual action, organizational change and global gender equality will be written by you.

References

[1] Brown, M. (2015). The Diversity Advantage: Why Hiring, Promoting and Funding More Women Will Boost Your Bottom Line. Accessed at: http://www.geekwire.com/2015/the-diversity-advantage-why-hiring-promoting-and-funding-more-women-will-boost-your-bottom-line/

[2] Wolley, A. & Malone, T. (2011). "Defend Your Research: What Makes a Team Smarter? More Women." *Harvard Business Review.* Accessed at: https://hbr.org/2011/06/defend-your-research-what-makes-a-team-smarter-more-women

[3] Catalyst. (2004). *The Bottom Line.* Accessed at: http://www.catalyst.org/knowledge/bottom-line-connecting-corporate-performance-and-gender-diversity

[4] Dezsö, C. L., & Ross, D. G. (2012). Does female representation in top management improve firm performance? A panel data investigation. *Strategic Management Journal, 33*(9), 1072–1089.

[5] Miller, C. (2016) "How Society Pays When Women's Work is Unpaid." *New York Times.* Accessed at: http://www.nytimes.com/2016/02/23/upshot/how-society-pays-when-womens-work-is-unpaid.html?_r=0

[6] Organization for Economic Co-operation and Development (OECD, 2014), "LMF2.5: Time Use for Work, Care and Other Day-to-Day Activities," OECD Family Database, Social Policy Division, Directorate of Employment, Labour and Social Affairs, (2014), http://www.oecd.org/els/family/LMF2_5_Time_use_of_work_and_care.pdf. The 2014 report uses data from 2008, the most recently available.

[7] International Labor Organization. *Maternity and Paternity at Work.* http://www.ilo.org/global/topics/equality-and-discrimination/maternity-protection/publications/maternity-paternity-at-work-2014/lang--en/index.htm

[8] http://www.telegraph.co.uk/women/womens-life/10572435/Intersectional-feminism.-What-the-hell-is-it-And-why-you-should-care.html

[9] Olsen, J. "Programs, Perceived Potential for Advancement and Organizational Attractiveness," *Group & Organization Management* (2015): 1–39.

[10] Al Dabbagh, M., Bowles, H. R. & Thomason, B. (2016) "Status Reinforcement: The Experience of Locals Negotiating at the Boundaries of Global Employment. *Working Paper.*

[11] Sandberg, S. (2013), *Lean In: Women, Work and the Will to Lead.*

[12] Slaughter, A. M. (2014), *Unfinished Business: Women, Men, Work, Family.*

[13] Ridgeway, C. (2011), *Framed By Gender: How Gender Inequality Persists in the Modern World.* Oxford University Press.

[14] Bowles, H. R., Babcock, L., & Lai, L. (2007). Social incentives for gender differences in the propensity to initiate negotiations: Sometimes it does hurt to ask. *Organizational Behavior and Human Decision Processes, 103*(1), 84–103.

[15] Uhlmann, E. and Cohen, G. (2005). "Constructed Criteria: Redefining Merit to Justify Discrimination," *Psychological Science* 16, no. 6: 474–80.

[16] Tetlock, P. E., & Mitchell, G. (2009). Implicit bias and accountability systems: What must organizations do to prevent discrimination? *Research in organizational behavior, 29*, 3–38.

[17] Staats, C. *State of the Science: Implicit Bias Review 2014* (2014), Kirwan Institute, Ohio State University.

[18] Frehill, Lisa M., *Why Do Women Leave the Engineering Workforce?*, (Society of Women Engineers: 2008).

[19] Fouad, Nadya, *Leaning In, But Getting Pushed Back (And Out)*, (American Psychological Association: 2008).

[20] Hochschild, A. (1997) *Time Binds.* Macmillan Publishers.

[21] Pew Research, *Modern Parenthood* (2013), Accessed at: http://www.pewsocialtrends.org/2013/03/14/modern-parenthood-roles-of-moms-and-dads-converge-as-they-balance-work-and-family/

[22] Wojcicki, S. (2014). "Paid Maternity Leave Is Good For Business", *Wall Street Journal*. Accessed at: http://www.wsj.com/articles/susan-wojcicki-paid-maternity-leave-is-good-for-business-1418773756

[23] Coltrane, S., Miller, E., DeHaan, T., Stewart, L. "Fathers and Flexibility Stigma," *Journal of Social Science Issues* 69, no. 2 (2013): 279–302.

[24] http://nytlive.nytimes.com/womenintheworld/2015/04/02/women-in-iran-post-videos-online-of-themselves-walking-in-public-with-their-heads-uncovered/

Afterword

Dave Ward is the CTO of Engineering and Chief Architect at Cisco Systems.

When building the strongest and highest performing engineering and innovation teams, seeking out diversity is a requirement. When reading that word, some will assume that I mean only gender or racial diversity, but I've always cast a much broader net, including experience, perspective, social and group compatibility, raw intelligence, ambition, energy level, inventiveness, engineering skills (type and level), problem-solving experience, and leadership talent, as some of the most heavily-weighted variables that make up diversity. Also, as a manager of top technical talent, I strive to expand the individual diversity of each of my team members by deliberately offering (or creating) experiences to help them reach the next levels in their careers. Depending on the individual, that could mean incremental or divergent growth within their field of engineering, a change in career trajectory, or even a complete self-reinvention. (Because I'm working with some of the brightest in the industry [and world], I see the complete reinvention more often than may be expected).

When hand selecting the individuals with the brightest and highest potential in the industry, my teams usually are naturally gender diverse. That said, in the networking industry, women who have reached the highest levels of the corporate technical ladder are few and far between, and those who have reached the pinnacle (usually labeled "Fellow") are almost nil. The CTO title is another, but this one tends to vary widely by organization and industry.

For those unfamiliar with these terms (setting aside the inherent lingual issue), there is a set of well-known corporate technical leadership career levels, just as there is for management leadership. The pattern is (with variations across the industry): Technical Leader (usually analogous to Manager or Senior Manager), Principal Engineer (similar to a Director title), Distinguished Engineer (analogous to Senior Director, and often with a role clarification like Sales Solution Engineer, Sales Engineer, Systems Architect, Technical Marketing Engineer, Consulting Engineer).

We talk so much these days about gender diversity in the tech industry, and we've made some good progress toward growing the pipeline of young female technical talent, particularly in computer science. While we have to keep the conversation alive, we also must place some real emphasis on experienced women in the technical workforce, and why they are hitting a glass ceiling on the technical career ladder. Over the past 15 years, many more women have been hired and promoted to levels of technical distinction within their companies, but far fewer than the number of men. "Women in Tech Leadership" stats are frequently published with good intent of celebrating progress, but those studies and stats rarely focus on senior technical positions or the technical career progression—that is, real female engineers who still do real engineering. (There is nothing wrong with a woman choosing the management track, and accomplishments in those ranks are still noteworthy, but here I am focusing on the women who want to grow their careers *as engineers*).

As a techie, and given my technical leadership role, I put an emphasis on innovation, solving customer problems, the advancement of technology, and the empowerment of engineering talent. I also try to keep up with all social science research related to tech talent, because it helps me to continuously expand my own strategies and tactics for attacking this issue. What disturbs me, though, is that there is almost no holistic social science research on gender diversity issues in the corporate technical career ladder. Most studies either generalize too much to glean potential root causes (lumping everyone and every role into "engineer"); specialize too much to be pragmatically useful in their application to engineers; or lack of p-value rigor to make concrete conclusions or comparisons across engineering disciplines. The lack of attention to the very real factors that make it difficult for women to advance, or *want to* advance through the technical ranks represents a massive gap in the state of social science research. I can't solve that gap in this piece. I can take a stab at illuminating it, and at enabling researchers in the field to look at this issue in earnest, based on what real women in real engineering roles already know with an eye toward root causes and solutions. I will also offer some of my own insights and experiences grounded in many years of managing technical women. I hope this will help to encourage both researchers and managers to keep at this problem with the same persistence they would put toward any other technical or research problem.

The techie path is, without a doubt, full of some of the world's most creative thinkers, and we all know that the "world's most creative thinkers" = "only men." Importantly, the industry's claim is that the tech career ladder is (or should be) a pure meritocracy. One's intellectual aptitude and ambitions should be the most important factors; and one's technical skills, technical leadership skills, and ability to execute should be the primary measures of evaluation for promotion.

Speaking about the industry at large, I can tell you that even the most basic gender diversity statistics—measuring the number of women with senior technical titles over time, throughout the technical career ladder—show that meritocracy has more noise than signal. I see similarly dismal results when looking at intellectual property statistics, industry group leadership positions, and authorship of technical specifications; and when searching job satisfaction, CV sites and lines of code committed in open-source projects. There are obviously much more sophisticated statistics we could measure and cite, but my point is that we are not even covering the basics on this issue.

Moving to a less theoretical topic, I want to also discuss some practical strategies for managing female engineers. Gender diversity in Silicon Valley (Si Valley) has become a hot topic in industry trade rags over the past ten years or so, and we have seen advice to managers (both women and men) about how to manage, engage, and retain women in tech. I find that most of this advice is too general to be useful for those managing the best and brightest female engineers. The mantras of *listen, talk, encourage, create work—life strategies*, and *create group diversity* are all well-known, can't be taken for granted, and must be constantly reinforced in managing both men and women. There are a few items specific to managing our geekiest women (and those who know me know that I only use that term with the highest affection and respect), that could help start the conversation. I'd like to share a few of those things, and a few techniques that I have found to work well in providing great working experiences (and advancements) for both accomplished and growing female engineers. Stating the obvious, every individual is unique, and any generalizations I may make or imply are just generalizations, but ones that I think may be useful to the cause and hopefully make my points. I'll leave it to our social scientist colleagues to prove or disprove their validity.

Let me say this bluntly: women are often not explicitly managed toward the most challenging engineering opportunities designed for the fastest possible advancement. That could be for a variety of reasons, some more or less obvious, which I won't try to capture here. It is a problem, though, and I've tried a couple of ways to fix it. As a technical leader, and manager of some of the industry's top networking talent, I see it as core to my role to work and talk with engineers about career opportunities, which for a techie really means "what's new, cool, hard, unsolved and really important." For an engineer to advance, he or she *MUST* be put on high visibility, high-value projects that have industry changing potential in a field or domain. In my domain, that means projects destined to change the Internet. I'm now going to say something a bit more subjective and experiential, but just as bluntly: for another set of reasons, some more or less obvious, women (on average) *seem* to be less likely to self-select for these high-visibility projects with industry changing potential.

The first half of my role as a manager of engineering talent is to match every engineer on my team (female or male) with the right problem to solve (or technology to invent) for the greatest mutual benefit of the individual and organization. In working specifically with female engineers, that means a few things. Sometimes, it means you have to listen a little harder and pay closer attention to your conversations and calibrate for cues. I've found that the women on my teams (again, painting with a broad brush) have tended to communicate less directly or decisively than the men about what they want, and I've had to spend more time teasing that out of conversation, testing and validating assumptions, and talking through options ahead of deciding on a course. When you spend the time to have those conversations, you get to know a person, and you begin to understand them. An engineer's unique talents, drives, motivators, and ambitions usually shine quite brightly when you know where to look for them. Once you find them, it is surprisingly easy to identify the right challenges and opportunities—but be careful not to underestimate your female engineers' abilities, because women also *seem* to have a greater propensity for understating their talents and skills. (Modesty is an admirable trait, but it can be very problematic if a manager is not paying attention).

The second, and most important, part of my role (after finding the right project) is to enable every engineer with the support structure necessary to succeed. Sometimes that means surrounding him or her with great teams. Sometimes it means enabling flexible hours. Sometimes it means scoping the project, or crafting the environment around a specific set of life circumstances. Every one of us has a giant sine wave (or pick your favorite curve) of creativity and productivity flowing through our brains and lives. We are all affected at different times by lack of sleep, life stress, impacts of commute, increases in responsibilities for kids or aging parents, relationship troubles, challenging geo-locales, and general burnout; or, conversely, mania, personal and family achievements, magical light bulbs going off in our heads, being touched by one of the gods, or those special shower moments in which so many of the industry's problems seem to have been solved. There are periods of massive productivity, and there are periods of creative drought. These things, too, have to be managed and grasped opportunistically by managers. Finding the right person—for the right problem or invention—at the right point in their life—and managing the project pace and stresses accordingly—can seem like an intractable problem. The key is being able to recognize (again, by talking with and knowing the person) when someone is on the sine wave, and to plan accordingly. This sounds obvious, and this management technique is rather generic, but it's also important to recognize that the waveform is sometimes a bit different in female engineers.

Women, more commonly than men, hold primary caretaker roles for children or aging parents. For anyone in a primary caretaker role, variable or flexible scheduling is key. I'm not stating anything new here, but there is a twist for techie women. In my experience, female engineers—particularly those balancing many work and home responsibilities—have been (perhaps out of necessity) nothing short of superb at time management. However, managers still need to be mindful of planning for due dates (e.g. code completion or shipping product). This can't be a shock to anyone. Stated more bluntly: if you're managing an engineer who is also a primary caretaker of school-aged children, don't put code complete during the kids' spring break or some other important holiday or event. It's not going to be successful. The meta-point here is, if you're paying attention, you'll know something about the family and social pressures that each of your employees face, and you'll plan accordingly. Primary caretakers at home (often women) can't tolerate management stupidity at work. To avoid management stupidity, you need to listen to and talk with your engineers, and work with them to create plans grounded in reality. When planning and forming teams, consider explicitly time zone compatibility, colocation requirements, and travel needs. Don't create impossible situations, and don't ever require a primary caretaker to choose between work and family. They'll self-select away from managers, teams, projects, and even their career if forced to.

Another challenge that women often face—especially when also balancing full plates at home—is that of reluctance (or lack of time) to proactively seek professional role changes. This challenge is nuanced, and is sometimes rooted in pragmatic realities (like the physics of time and space), but can also be rooted in unfounded (but understandable) fears. Let me explain.

Most engineers have a full understanding that, if you own a horse, you have to clean out the stables. Translated, there are non-super-fun jobs as an engineer, but everyone knows that comes with the turf. One can easily become stuck in the bug-fixing trenches, forklifting of code, integration hell, or grinding out of well-trodden specs and standards for the umpteenth time, and never be given a chance to break out. If not highly proactive and vocal, an engineer can easily become pigeon-holed, which can cause career stagnation, intense boredom, or worse.

For unknown reasons, over the years, I have found many brilliant women tucked away in the trenches, doing the hard labor, and who, once invited to the cast, have become stars in innovating technology, tech strategy and in driving tech projects. When I've asked them why not sooner, I have heard variations on a theme—too busy to even think about it, fear of making time commitments they can't meet due to their family obligations, fear of disrupting their job stability (a particular sensitivity for single moms), fear of getting into a role where they don't yet have the skills to perform to their own standards, and fear or guilt that leaving their current job undone (or not done as well by someone else) will result in reverse progress or angst for their current teams, where they hold great loyalties. These fears keep even the smartest women from vocalizing their ambitions, and from self-selecting for great opportunities that could put them heads and shoulders above many of their male counterparts. I don't know how to solve this problem at a root cause level, but I do believe it's every manager's responsibility to seek these women out, understand their talents and ambitions, and pull them out of the quicksand. These are the women who should be filling your top technical talent pipelines, and for reasons likely rooted in larger societal issues, they are pushing through the muck, cleaning the stables, and never complaining, afraid of the unintended consequences that change may bring. Often unobservant managers only learn of these issues when the badge is on the table and she's walking out the door.

Leaders of large corporations often wonder why it is particularly difficult to retain female engineers. Getting caught in non-advancing roles, not being rewarded for hard work, and not being selected for or finding a way into new opportunities, are perhaps the top reasons why engineers leave their jobs. If you couple those factors with the common challenges that women face, particularly in large "traditional" organizations, and the basic human behavioral *principle of least effort* (or *path of least resistance*, if you're an engineer), it's not difficult to imagine why so many female engineers leave large corporations about midway through their technical careers. Further, Si Valley can be a badge-swapping extravaganza in its own right. Some believe the only chance for advancement is to leave one company for another (and maybe come back later, usually at a much higher level or pay grade). I firmly disagree with that philosophy, but theoretical disagreement and corrective action are two different things, and this very real problem is arguably magnified for women who face glass ceilings and unconscious biases at so many different levels in the tech career ladder. I've spent much of my career thinking about and trying to solve this problem.

For decades, I've been fielding a perpetual line of engineers at my door who are willing to work nights and weekends, and any other spare moments, on cool "stretch" (also commonly known as "20%") projects, just to be given the opportunity to use all their creativity as engineers. I spend an abundance of my own time as a positive innovation catalyst, trying to connect these engineers—men and women—with the right problems and opportunities—at the peak of their sine waves—and to help them (and their managers) create the right frameworks and teams for their success. I will probably still be fighting that battle for as long as I'm in the industry, but it's a worthy one, and I hope that other managers (or future managers) reading this will find a way to create a system around this tenet. It also would be very interesting for social scientists to analyze these and other positive (and negative) catalysts of innovation and gender bias disruption.

My last point for managers (though I could go on, this *is* just the Afterword) is this: jerks and bullies can't be tolerated on any team. Engineers are, by nature, argumentative and competitive. That is a good thing. The willingness to bullishly pursue excellence, sometimes at great cost, is how the best ideas and solutions are born. The theory is that the best idea wins, though it's not always that simple, and it often means that the best-*defended* idea wins, or that the strongest defender of his or her idea wins. The ability (and willingness) to argue for what is "best," and conversely defeat what is "not best," can make for a challenging environment for anyone, and is not for the faint of heart. Best is often an argumentative expression of math, design, physics, coding skills/style, understanding the power of X or Y programming language, performance, scaling, or algorithmic prowess. The best dialectic usually wins, and trust me— you've never seen a strutting display of dialectic feathers quite like that of a male software engineer whose code just (finally) compiled, thus solving the universe's most difficult and important problem. (Just ask him.) Given this environment, and a frequent lack of social graces on the parts of our male geek counterparts (sorry guys—let the social scientists prove me wrong), the end result is often that those who aren't "aggressive" (or willing to employ aggressive techniques) don't "win" the arguments, and thus don't fare as well in the technical Darwinism that exists in most corporations. It's the world of techno-politics (yes, engineers have their own political constructs and "argumentative athletics" = cardio). Conversely, those who pursue that categorization of "best" at the expense of other (potentially better) ideas, or of their teammates—often discussed in social narrative as those having been told throughout their lives and careers that they were "special"—can simply destroy the culture of a team, and, in particular, a team with women on it.

It's your role, as a manager, to lead those discussions and meetings, and to ensure that the best ideas are always heard, and always prevail. Make sure the engineers on your team who are less wired to argue do not check out—this is a surefire way to miss some of the best ideas. While female engineers have been known to *not* self-select for some of the best career opportunities, they do tend to self-select themselves *out* of harsh, jerk-laden, feather strutting environments that put their advancement (and sanity) at risk. They also tend to veer away from leadership roles that would put them at the helm of an already hostile team culture. No one likes having a jerk, bully, or social tumor on their team. Manage your team such that everyone can express their skills fully; and jettison or isolate engineers (even the best ones) who are poisonous to the team culture you're trying to create.

So many engineers have gotten away with bad social behavior because they were brilliant. The fact that people refuse to work with them should be a clue that in the ROI of that brilliance is team malfunction, and the return is negative. If any engineer tries to lay claim to (or be anointed as) "always having the best idea," confiscate his (or her) badge immediately, and don't give it back, for one who lays that claim has almost certainly lost their objectivity and clearly their ability to recognize their own strengths and weaknesses.

If you are still reading this, and you believe what I wrote in the last paragraph, you'll realize the importance of communication skills for engineers pursuing tech leadership roles, and for the managers of those engineers. There's a ton of educating, facilitating debates, and communication about ideas, problems, solutions, code, designs, new skills, specs, and written/presentation work that is just part of the job. Most engineering curricula in universities don't weight (or teach) these skills as heavily as they are used in practice, and thus don't adequately prepare engineers for the working world. Unfortunately, most corporate development programs for engineers are also woefully inadequate in this area. I wish I had known this when designing the first parallel processing, multi-threaded operating system for a distributed router. I assumed that because it was a long-standing topic in computational operating system engineering, every engineer in the networking industry would already know the concepts. I hired the best coders I could find and almost failed. I should have hired the best educators and communicators I could find. The best coders in the field couldn't communicate and teach the new concepts to the others, and that put the project at risk. I had to retrofit the construction of the team and the design-patterns we utilized and take on the education of the team myself. Back to the topic at hand, most programs that claim to tackle diversity, leadership, and public speaking, in fact, miss some major categories of skills that engineers need (including dialectic, technical communication, and some of the social graces I alluded to above), and thus just completely miss the mark for a geek or geek-leader. The one-size-fits-all programs we regularly see are, unfortunately, part of the Dilbertian stigma that has risen in Si Valley.

The problem of big initiative programs affects both men and women, but I've found that one way to overcome this for women (at least in part), is to create a grass-roots technical community, led by women, where engineers can talk tech with each other without the typical Si Valley burden of feather-fluffing, mindless arguments over nits and moot points, and endless self-gratification. Six or so years ago, one of the women with whom I had been working with for more than a decade came to me with an ultimatum: "I need to do something more challenging, or I'm going to lose my mind, or leave the company. I'm willing to work for you because you think you can help, but if you can't fix this problem, I'm gone." No good engineering manager can pass up a meaty problem like that, so I took her up on the challenge. Solving the problem for her was low-risk for me, as this woman is brilliant, and was one of the most influential young women during the formation of the Internet. It so happens that she, like many other mid-career women, had found herself cleaning out the engineering stables for way too long, and she was ready to kick ass again. I found her a great (and really hard) problem to solve—one that fit well with her skills and sine wave—and gave her something completely new to think about and learn. I surrounded her with great engineers, connected her with interested customers, and she has since cranked out several industry changing products and technologies, and is still going strong.

Back to my point about communities: I told her that I would be willing to take a crack at the larger problem, too, but only with her help. I asked her to form a women's engineering group. She refused. "No way," she told me. "I'm not hosting yet another group meeting where we share our feelings about what it's like to be a woman. We *are* women—we already know what it's like." It took a few years, but I eventually convinced her to give it a shot by planning a single, covert instance, under the corporate radar. She replied, "I'm only going to do this if there's a technology component, and if we have female geeks as speakers, presenting their technologies or projects—not their feelings," she said. Fortunately, that was exactly what I wanted, and thankfully, one meeting led to another, and her forum has now become one of the most successful communities in the company, and in the Valley. Over 100 women in the company join every month to geek out on a new technology topic. The community is (proudly) still running completely under the corporate radar, and the word has spread. We now have women and men from across Si Valley attending and presenting regularly. Providing female engineers opportunities to present (with obvious pride) their *engineering* accomplishments, to take questions, to teach, listen, and learn in an environment without the lens of the big corporate eye, were more in demand than we ever thought, and it was the pure will and persistence of my friend—one of the greatest female engineers and leaders with whom I've had the pleasure of working—that has made this community a success. A program by the female geek, for the female geek took a long time to create, but has paid off in huge dividends, and has helped me (and my peers) to identify many of those women who are stuck in quicksand, and to pull them out. It has also become a positive institution in our culture, and has attracted geeks of all genders. True geeks, after all, really just want to geek out, and we all need to get back to those roots.

Throughout this book, there are many inspiring stories by women who have built successful technical careers, in part due to, or at times in spite of, the management they have received along the way. These stories are designed to inspire, and not all of them will expose the unique challenges they've faced, either at work or at home. I know many of these women personally, and would encourage you to reach out to them with your own stories, or to ask them directly about some of the hurdles they've had to overcome, including the few I've presented. If there is one common theme I would like you to take away from my anecdotes, and from the stories in this book, it's that of persistence. These women have been persistent, and they have sought out managers and mentors who have been persistent in helping them to succeed. If you are a woman engineer, you have almost certainly run up against many of the challenges I mentioned—or you may—but you must not give up. Continue to work your ass off. Make time for a stretch project. Find another manager, or join another project, if you must. Find another company, if you must. But don't give up on our industry. It's our responsibility, as engineers, to solve this problem.

So why, in this short Afterword of a book written by people much more qualified than me to write about these subjects, have I gone into such great depth about the problems facing women in engineering? In my industry, we've built the technology underpinning the Internet with a fundamental philosophy of making it available to everyone, everywhere. The technology is not gender-specific; nor is its design. That said, the social factors surrounding the ways in which technologies are built is a subject area of great importance in social science, and one that is woefully under-analyzed. Gender studies miss some of the most basic blocking and tackling, and are often not designed with root cause analysis in mind. This fundamental gap in our research underserves the topic of women in engineering. We have yet to answer honestly some of the most foundational questions, like: What are the diversity statistics throughout the technical career ladder, and over time? What are the root causes of ceilings or stagnation for women at each level in the ladder? Which of those causes are environmental, and can be controlled by better management?

What is the impact of stresses, either at home or at work, on the creativity and productivity sine waves? These are just a few areas worth investigation. We need data beyond the typical simplistic summary stats, and we need to get to root cause if we are going to sustainably solve these problems for the engineering community. We need to understand deeply the unique challenges of women navigating technical careers, and we need to hire and groom technical managers who know how to enable female engineers to move up in the technical ranks. We need to expand the honest conversation about male advocacy, gender related communications, engineering management, and education; and we need to tailor our corporate leadership programs to be more specific to the technical ranks. Techniques for managing engineers have to mature. We are at a critical point of inflection in the gender diversity conversation, and if we fear the open and candid conversation, we run the risk of diluting or worse, reversing our progress. We have a responsibility, as engineers, to attack the problem head on, and to solve it for the engineering and wider community.

For those who see value in the experiences I've shared, please take the time to work some of those ideas into your own professional lives. For those social scientists or researchers who feel compelled by the implied hypotheses, please take the time to prove or disprove them. For those who feel that I've overgeneralized or put forth unfounded claims from extrapolation of personal anecdotes proved by emphatic assertion; please take the time to debunk them. It's time to be honest about the (lack of) data, and to make a concerted effort at fundamentally improving the landscape for our female engineers who wish to attain the highest technical ranks.

List of Contributors

Rahilla Zafar, Lead Editor and Author

Rahilla Zafar was contracted by Cisco Systems while researching, writing, and editing this book. She traveled to over seven countries conducting field research including Saudi Arabia, Jordan, Israel, Palestine, and the United Arab Emirates. Rahilla also helped source and edit content for this book working with contributors from over 30 countries. In August 2016, she joined the leadership team of ConsenSys, a New York-based global venture production studio, building decentralized applications on the Ethereum blockchain. Rahilla has also consulted for start-ups and non-profits including Futurism and the EurAsia Foundation. She previously was based in Afghanistan and Pakistan working for the International Organization for Migration, the NATO-led mission, and government initiatives focused on humanitarian outreach. She holds two graduate degrees from the University of Pennsylvania and the London School of Economics and received her undergraduate degree from DePaul University. While a graduate student at the University of Pennsylvania, Rahilla co-authored *Arab Women Rising* with the award-winning journalist Nafeesa Syeed highlighting women entrepreneurs from across the Middle East published by *Knowledge@Wharton (K@W)*. She's been a contributor to *K@W* with a long-time focus on innovative technologies and its impact on women in particular. She also was a contributing writer for *K@W* interviewing participants and writing pieces for the Goldman Sachs 10,000 Women Initiative.

Nada Anid Ph.D., Co-Editor and Author

Nada Marie Anid, Ph.D., is the first female dean of New York Institute of Technology's School of Engineering and Computing Sciences. In this role, she oversees over 80 engineering and computing sciences faculty members and 3,500 students in Manhattan, Old Westbury, N.Y., Abu Dhabi, Vancouver, and Nanjing, China.

Dr. Anid embraces NYIT's forward-thinking and applications-oriented mission and is working on several strategic partnerships between the School of Engineering and the public and private sectors, including creating the school's first Entrepreneurship and Technology Innovation Center (ETIC) and its three labs in the critical areas of IT and Cyber Security, Bioengineering, and Energy and Green Technologies.

She is committed to tackling the challenges identified by the White House Strategy for American Innovation and the National Academy of Engineering. Long an advocate for women pursuing career opportunities in engineering and technology, Anid is the recipient of numerous awards in recognition of her business acumen and her support of diversity and women in technology. She was one of the winners of the 100 Inspiring Women in STEM Awards by *Insight into Diversity* magazine and of the LISTnet Diamond Award in recognition of her significant contributions toward the advancement of women in technology, as well as for her professional achievements in the technology field. Anid was also three times named one of the top 50 women in business on Long Island by *Long Island Business News* (LIBN) in recognition of her business acumen, mentoring, and community involvement, and was inducted to the LIBN Hall of Fame.

Anid holds leadership positions in the American Institute of Chemical Engineers, the American Society for Engineering Education, and the New York Academy of Sciences, among others. She earned her Ph.D. in environmental engineering from the University of Michigan (Ann Arbor), and BS/MS degrees in chemical engineering from the Royal Institute of Technology (KTH-Stockholm).

Laurie Cantileno, Co-Editor and Author

With many years in the industry, Laurie's hallmark is strategic technology leadership focused on innovation and execution, growing revenue opportunities, identifying tactical and strategic plans, building delivery focused organizations, driving collaborative partnerships, and advising C-level executives. Most recently, the Service Delivery Executive for Cisco and formerly as Vice President of Global Network Services at Broadridge Financial, Laurie is directly responsible for designing corporations' overall operational technology management framework, developing key relationships to drive multi-million dollar deals, creating the firm's first PMO, implementing a global network engineering structure as well as the creation of a business-focused collaboration solution internally as well as for her external clients. Tactically and strategically these business solutions not only saved these businesses operating expense, but dramatically reduced business risk as well as introduced continuity never seen before. Laurie has a track record of broad cross-functional leadership experience across all core business functions with emphasis on sales, finance, human resources, and productivity and performance improvement.

Laurie was the catalyst for the partnership between NYIT and Cisco resulting in the creation of The IoW and the collaboration on this book. This collaboration between Cisco and NYIT does not end there with the additional focus on the creation of Cisco Services Academy globally and Cisco's participation in the Annual NYIT Global Cybersecurity conferences in NYC and Abu Dhabi.

Laurie continues this leadership by providing guidance and mentoring to girls, young women and colleagues as a former Board Member of Girls Inc, former Broadridge Inclusion & Diversity Board Member and the Senior Women's Leadership Forum, a member of the Executive Technology Advisory Board for the NY Institute of Technology, as Advisor for the Society of Women Engineers and as the Leader at Cisco for the Global Enterprise Theater for Inclusion and Diversity. Additionally Laurie has created an annual "STEM

Rising Star" scholarship at NYIT for an exceptional junior-level female scholar financially in need and pursuing a career in the sciences. Laurie can also be found donating as Associate Producer on other STEM initiatives like the film *She Started It* featured in this book. The National Association of Professional Women (NAPW) also selected Laurie as Business Woman of the Year in 2014.

Her personal interests include organic gardening and fitness as a certified personal trainer and a group fitness instructor. One of her passions is speaking with and coaching young women about confidence and overcoming fear while pursuing their dreams. She is married and lives between Florida and NY with her husband and many animals. She holds a Magna Cum Laude degree in Computer Science from NYIT as well as an ITIL v3 certification for Service Delivery.

Monique Morrow, Co-Editor

Monique Morrow is the CTO of New Frontiers Engineering at Cisco. Focused on the intersection between economics, technology, and research, she is defining mechanisms and marketplace scenarios for cloud federation constructs to include security. She was previously Cisco's Chief Technology Officer of Services, where she was responsible for aligning the Cisco Services Technology vision and architectures with the business strategy.

Monique has a track record of co-innovating with customers, developing solutions that have transcended the globe from North America, Europe and Asia. Under Cisco's Office of the CTO, both as an individual contributor and manager, Monique built a strong leadership team in Asia-Pacific. Her specific geo-area targets were China and India. Monique's role in these regions drove Cisco's globalization and country strategies and met all of her targeted goals. Monique has consistently demonstrated the willingness and courage to take risks and explore new market opportunities. These innate qualities are part of her DNA and are of great value to Cisco and all the global organizations with which she is involved.

Monique was also listed as Top Ten Influential IT Women in Europe in 2014. She was one of 6 Global Achievers recognized for the ITU and UN Women GEM-TECH Award in 2014. In 2014 she was further recognized by IEEE Region 8 with the Clementina Saduwa Award.

In 2015, Monique was selected to be part of the elite group of women for *Connected World* magazine's 2015 Women of M2M/Internet of Things feature—or "WoM2M." Monique additionally listed as one of the top 50 most inspiring women in Europe for 2016. Monique has also been published in IEEE and other journals and speaks frequently at conferences; and has co-authored six books.

Susan Ann Davis, Foreword Author

Susan Ann Davis, a pioneering woman business owner, has grown Susan Davis International over three decades into a global strategic communications agency. Internationally known for her expertise in strategic positioning, reputation management, crisis and cyber risk communication, Ms. Davis provides counsel to key industry and government executives worldwide. A lifelong advocate for social entrepreneurship, democracy building, and leadership development for women, she is the Board Chair of Vital Voices Global Partnership and was the first international president of the International Women's Forum, now comprising over 7,000 women leaders. She is also the Board Chair of Razia's Ray of Hope Foundation supporting a school for 600 girls in Afghanistan, and plays a key board role with Medical Missions for Children, Self-Help Africa, the Ireland Fund America and the United States-Panama Business Council, among others. A graduate of University of Wisconsin-Madison, she is a recipient of the university's Distinguished Alumni Award.

Bobbi Thomason Ph.D., Closing Author

Bobbi Thomason is a Senior Fellow and Lecturer in the Management Department at the Wharton School of the University of Pennsylvania and a Research Fellow at the Women and Public Policy Program at the Harvard Kennedy School. Her research focuses on women's attainment of leadership and diversity within global organizations. Her dissertation explored the career paths of female executives in the Middle East and North Africa. Bobbi earned her BSFS from Georgetown University, her MA from Columbia University, and her Ph.D. from Stanford University.

Dave Ward, Afterword Author

Dave is Chief Architect at Cisco Systems, responsible for architectural governance, defining strategy, development of new technology and leading use-inspired research. Working via tight partnerships with customers, partners, developers and academia he is also leading co-development and co-innovation initiatives. He has been the Routing Area Director at the IETF and chair of four Working Groups: IS-IS, HIP, BFD and Softwires and worked with the ITU-T, ONF and several open source consortia. David was also a Juniper Fellow and Chief Architect working on the operating system and next-generation routing systems. Dave has a small vineyard in the Santa Cruz Mountains, and an heirloom tomato farm along the St. Croix River in Somerset, Wisconsin.

Azam Zafar, Research Editor

Azam Zafar is a digital product designer and customer experience strategist. His experience spans from advising early stage startups to enterprise consulting, working with clients such as Salesforce.com, Emirates Airline, and Citibank. He holds a Bachelor of Science with honours in Government from the London School of Economics and is currently pursuing postgraduate studies in the department of Computer Science at the University of St Andrews in Scotland with interests in Human-Computer Interaction and Artificial Intelligence.

Contributing Authors

Aglaia Kong
CTO, Corporate Network, Google
San Francisco, CA, USA

Dr. Alison Vincent
Chief Technology Officer, UK & Ireland
Cisco International Ltd
UK

Alphonsine Imaniraguha
Network Consulting Engineer
Cisco Systems
Raleigh, NC, USA

Amal AlMutawa
Senior Project Manager
Prime Minister's Office – UAE
Dubai, United Arab Emirates

Ameen Soleimani
Software Engineer
ConsenSys
New York, NY, USA

Andrea Barrica
Venture Partner/Entrepreneur in Residence
500 Startups
San Francisco, CA, USA

List of Contributors 343

Ankur Kumar
Director of Programs
McKinsey Academy
New York, NY, USA

Aurore Belfrage
Head of Together
EQT Ventures
Stockholm, Sweden

Ayanna Terehas Samuels
Development Consultant specializing in ICTs and Gender-related issues and Technology Policy Expert
Self-employed
Kingston, Jamaica

Banu Ibrahim Ali
IT – Specialist
Slemani, Kurdistan

Brian Rashid
Professional Speaker, Writer, and Storyteller
New York, NY, USA

Cindy M. Cooley
Program Manager
Cisco Systems
Silicon Valley, CA, USA

Dimitri Zakharov
CEO
Impact Enterprises International
New York, NY, USA
Chipata, Zambia

Edna Conway
Chief Security Officer, Global Value Chain
Cisco Systems, Inc.
New Hampshire, USA

Fereshteh Forough
Founder and CEO
Code to Inspire
Herat, Afghanistan
New York, NY, USA

Iffat Rose Gill
Founder and CEO
ChunriChoupaal – The Code To Change
Netherlands/Pakistan

Iliana Montauk
Director
Gaza Sky Geeks, Mercy Corps
Gaza

Karoline Evin McMullen
Senior Strategist
iconmobile
New York, NY, USA

Laila Abudahi
Fulbright Scholar
Firmware Engineer, Palo Alto Networks
Gaza

Lauren Cooney
Sr. Director, Strategic Programs, Chief
Technology and Architecture Office
Cisco Systems
San Jose, CA, USA

Martin Lundfall
Software Engineer
ConsenSys
New York, NY, USA/Stockholm, Sweden

Mary Aiken
Director
Cyberpsychology Research Network
Ireland

Meltem Demirors
Director
Digital Currency Group
New York, NY, USA

Miriam Grobman
Founder and CEO
Miriam Grobman Consulting
Austin, TX, USA

Mitali Rakhit
Founder and CEO
Globelist
New York, NY, USA and Dubai, UAE

Noa Gafni
CEO and Founder Impact Squared
London, UK, and New York, NY, USA

Sarah Judd Welch
CEO/Head of Community Design
Loyal
New York, NY, USA

Sarah K. Thontwa
Researcher
International Food Policy Research Institute
Kinshasa, DRC

Shraddha Chaplot
Greengineer
Cisco Systems
San Francisco, CA, USA

Tahira Dosani
Managing Director
Accion Venture Lab
Washington, DC, USA

Tess Mateo
Founder and Managing Director
CXCatalysts
New York, NY, USA

Tracy Killoren Chadwell
Partner
1843 Capital
Greenwich, CT, USA

Tracy K. Saville
Co-Founder and CEO
Sofiia AI
SRI/Menlo Park, CA, USA